Lecture Notes in Artificial Intelligence 8313

Subseries of Lecture Notes in Computer Science

LNAI Series Editor

Randy Goebel
University of Alberta, Edmonton, Canada
Yuzuru Tanaka
Hokkaido University, Sapporo, Japan
Wolfgang Wahlster
DFKI and Saarland University, Saarbrücken, Germany

LNAI Founding Series Editor

Joerg Siekmann
DFKI and Saarland University, Saarbrücken, Germany

Jordi Nin · Daniel Villatoro (Eds.)

Citizen in Sensor Networks

Second International Workshop, CitiSens 2013
Barcelona, Spain, September 19, 2013
Revised Selected Papers

 Springer

Editors
Jordi Nin
Barcelona Supercomputing Center (BSC)
Universitat Politècnica de
  Catalunya-BarcelonaTECH
Barcelona
Spain

Daniel Villatoro
Barcelona Digital Technology Centre
Barcelona
Spain

ISSN 0302-9743   ISSN 1611-3349   (electronic)
ISBN 978-3-319-04177-3   ISBN 978-3-319-04178-0   (eBook)
DOI 10.1007/978-3-319-04178-0
Springer Cham Heidelberg New York Dordrecht London

Library of Congress Control Number: 2013957356

CR Subject Classification (1998): I.2.9, H.2.8, I.2.6, I.2.11, H.3.4-5, H.4.1-3, H.5.3, J.2

LNCS Sublibrary: SL 7 – Artificial Intelligence

Printed on acid-free paper

Springer is part of Springer Science+Business Media (www.springer.com)

Preface

The current volume constitutes the revised proceedings of the Second International Workshop on Citizen Sensor Networks (CitiSens 2013), which includes revised versions of the papers presented at the workshop. The aim of CitiSens is to promote and stimulate the international collaboration and research exchange on novel smart cities and sensor networks topics. This second edition of the workshop was co-located with the ECCS 2013 conference in Barcelona (Catalonia, Spain).

The program of this year's workshop consisted of the presentation of seven accepted full papers (out of 16 submitted papers). The accepted papers deal with topics such as trajectory mining, smart cities, multi-agents systems, networks simulation, smart sensors, clustering, or data anonymization. Each paper was reviewed by at least three reviewers.

We would like to acknowledge and thank all the support received from the Program Committee members, external reviewers, and the Organizing Committee of ECCS 2013. We would like to warmly thank Josep Domingo-Ferrer for his support through the UNESCO Chair in Data Privacy.

Last, but definitely not least, we would like to thank all the authors who submitted papers, all the attendees, and the keynote speakers, Dirk Helbing (ETH Zurich, Switzerland) and Daniele Quercia (Yahoo! Labs), who accepted our invitation to give two talks, entitled "Sensorship or Censorship - That's the Question" and "Crowdsourcing for the Good of London," respectively, for all the attendants of Citisens 2013.

December 2013

Jordi Nin
Daniel Villatoro

Second International Workshop
on Citizen Sensor Networks – Citisens 2013

Organizing Committee

Jordi Nin   Barcelona Supercomputing Center (BSC), Universitat Politècnica
                   de Catalunya-BarcelonaTECH, Barcelona, Spain
Daniel Villatoro   Barcelona Digital Technology Centre, Spain

Program Chair

Jordi Nin   Barcelona Supercomputing Center (BSC), Universitat Politècnica de
             Catalunya-BarcelonaTECH, Barcelona, Spain

Scientific Committee

Andrea Baronchelli	Northeastern University, USA
Ana Bazzan	Universidade Federal do Rio Grande do Sul, Brazil
Sergio Borger	IBM Research, Brazil
Longbing Cao	UTS, Australia
Francesco Calabrese	IBM Research, Ireland
Licia Capra	University College London, UK
Enrique Frias-Martinez	Telefonica Research, Spain
Jordi Herrera-Joancomarti	Autonomous University of Barcelona, Spain
Fernando Koch	IBM Research, Brazil
Neal Lathia	University of Cambridge, UK
Ed Manley	UCL Casa - Center for Advanced Spatial Analysis, UK
Yamir Moreno	Institute for Biocomputation and Physics of Complex Systems, Spain
Mirco Musolesi	University of Birmingham, UK
Alexei Pozdnoukhov	NCG- NUIM, Ireland
Daniele Quercia	Yahoo! Research, Spain
Sergio Ricciardi	UPC-BarcelonaTECH, Spain
Savaltore Rinzivillo	KDD-Lab, ISTI - CNR, Italy
Victor Rodríguez Doncel	Technical University of Madrid, Spain
Francesco Ronzano	Barcelona Digital, Spain
Frank Schweitzer	ETH Zurich, Switzerland
Jetzabel Serna	Barcelona Digital, Spain

Contents

Trajectory Mining

Actions in Context: System for People with Dementia

Àlex Pardo[1]([✉]), Albert Clapés[1,2], Sergio Escalera[1,2], and Oriol Pujol[1,2]

[1] Dept. Matemàtica Aplicada i Anàlisi, UB, Gran Via 585, 08007 Barcelona, Spain
[2] Centre de Visió per Computador, Campus UAB, Edifici O, 08193 Barcelona, Spain
alexpardo.5@gmail.com, aclapes@cvc.uab.cat, sergio@maia.ub.es,
oriol_pujol@ub.edu

Abstract. In the next forty years, the number of people living with dementia is expected to triple. In the last stages, people affected by this disease become dependent. This hinders the autonomy of the patient and has a huge social impact in time, money and effort. Given this scenario, we propose an ubiquitous system capable of recognizing daily specific actions. The system fuses and synchronizes data obtained from two complementary modalities - ambient and egocentric. The ambient approach consists in a fixed RGB-Depth camera for user and object recognition and user-object interaction, whereas the egocentric point of view is given by a personal area network (PAN) formed by a few wearable sensors and a smartphone, used for gesture recognition. The system processes multi-modal data in real-time, performing paralleled task recognition and modality synchronization, showing high performance recognizing subjects, objects, and interactions, showing its reliability to be applied in real case scenarios.

Keywords: Multi-modal data Fusion · Computer vision · Wearable sensors · Gesture recognition · Dementia

1 Introduction

The number of people living with dementia increases every year. According to The World Alzheimer Report, in 2010 there were 35.6 million dementia affected people and this number is expected to increase to 65.7 million by 2030 and 115.4 million by 2050. The most common type of dementia is Alzheimer's disease. It mainly affects elder people and the most common symptom of this disease is lack of awareness. This hinders the autonomy of the patient because he could not remember where he is, what he had already done, or what he has to do. As a result, he requires constant attention. Intelligent systems help patients to reduce their dependence on carers and improve their quality of life in the early stages of the disease. In this paper, we present an ubiquitous system to assist people with dementia. Our framework can be split into two main modalities, an "ambient" one based on multi-modal visual data for user, object, and user-object relationship recognition, and an "egocentric" one based on wearable sensors to model user gestures.

J. Nin and D. Villatoro (Eds.): CitiSens 2013, LNAI 8313, pp. 3–14, 2014.
DOI: 10.1007/978-3-319-04178-0_1, © Springer International Publishing Switzerland 2014

From an ambient perspective, different visual-based sensors are placed in the environment and computer vision and pattern recognition techniques are applied. In this field, different ubiquitous systems have been recently proposed in order to assist people with dementia, most of them focused on the detection of falls and risky events. In [2], the authors present a privacy preserving automatic fall detection method. The system is able to recognize five different scenarios (standing, fall from standing, fall from chair, sit on chair, and sit on floor) based on the 3D depth information provided by a 3D camera. Whereas the former system uses quite simple features, more refined approaches to the fall detection have been proposed. In [3], the authors use fuzzy clustering techniques to accurately detect the activity (sitting, being upright, or being on the floor) performed by an elder. In [1], a system for passive fall risk assessment in home environments using Microsoft® Kinect™(RGB-Depth camera) is presented. The risk is evaluated obtaining measurements of temporal and spatial gait parameters. Most of these works are based on background subtraction and people tracking techniques [8]. And, with the increasing adoption of depth sensor data, novel multi-modal RGB and depth object descriptors are being proposed [5,6].

From an egocentric perspective, previous works on wearable sensors and health-care are based on activity recognition for promoting active lifestyles and prevent related diseases [9,10]. The authors of [12] propose a Smart Reminder of intended activities for people with dementia. In this case, Dynamic Time Warping (DTW) [13] algorithm is applied to perform activity recognitions. DTW is also applied in [14] in order to perform context recognition by means of data acquired from wearable sensors.

Finally, few works have been recently presented combining the power of ambient and egocentric paradigms within the same framework. In [16] a single wearable sensor and a blob-based vision system is used in order to identify daily activities, such as walking, sitting, standing, laying down. The work shows improved recognition results when both modalities are jointly considered. In [17], bicycle maintenance activities are recognized using an ultrasonic hand tracking system and motion sensors placed on the arms. In [18], the authors use a camera worn by the user and an inertial sensor fused data to estimate the pose and create mixed reality scenes. Nevertheless, most of previous works use high complexity algorithms to fuse the data, such as Neural Networks, Hidden Markov Models [19], Gaussian Mixture Model Bayes classifiers [16] and Extended Kalman Filter in [18]. In those cases, the complexity of the algorithms make the reliable and non-invasive implantation of a real-time home-care ubiquitous system for assistance for the elder and people with dementia difficult.

In this paper, we propose a multi-modal fusion pipeline for real-time user detection, object detection, object identification, and user-object relationship recognition using computer vision by analyzing data from a RGB-Depth camera (Microsoft® Kinect™) combined with gesture recognition by means of processing streaming data from a 9-degrees-of-freedom wearable IMUS sensors (Shimmer®) which measure acceleration, angular speed and magnetic field strength and direction. In order to recognize the subject environment (context), the computer vision part models a RGB-D environment learning a Gaussian distribution for each pixel from a set of initial frames. Then, in each new frame, the foreground is segmented according to the confidence with respect to the learnt environment. From the depth image, subjects are also detected and tracked using robust depth features and a Random Forest classifier. The previously segmented foreground regions, not classified as subjects are considered to be objects. These object regions are described, matched, and recognized using 3D descriptors of normal vectors distributions (FPFH). In the Personal Area Network scenario (PAN), actions are recognized by means of a parallel proposed version of Dynamic Time Warping over the spatio-temporal measurements given by the wearable sensors. Thus, given an interaction between a patient and an object, we can determine which action is done by fusing both modalities. The proposed system is simple yet efficient, runs in real-time and has a high performance rates. The presented ubiquitous system is successfully evaluated on real multimodal data simulating real use-case scenario for people with dementia taking a medication.

The rest of the paper is structured as follows: Sect. 2 explains the proposed system. Section 3 describes the data, scenarios, settings and validation measurements and the results. Finally, conclusions are given in Sect. 4.

2   System

In this section, we present our multi-modal activity recognition system for people with dementia. Actions are meaningful with respect to the context where they took place. Actions in Context means mixing ambient and egocentric features, giving the system a two-sided reality, one from the perspective of the environment and the other from the point of view of the user.

Figure 1 shows the architecture of the proposed system. In the egocentric computing part, we have an IMUS Sensor connected with an smartphone using a Bluetooth® connection. This mobile device is used as a HUB and its function is to label every sample with a time-stamp and stream it using IP to the server which will process all the data in real-time. This part of the application is relative to subject who is wearing the inertial sensors. The ambient part consists on a RGB-D camera monitoring the scene. This device gives us information about color and depth. Processing this information allows us to be able to perform user and object detection and recognition. Both data streams are synchronized by means of the NIST Internet Time Service. The different modules of our system are described in detail next.

2.1   Egocentric Computing

This part of the system is centered on the gestures the user is performing with his/her dominant arm (left part of block shown in Fig. 1). For this task, we first compute a set of inertial features which are then used to perform gesture recognition based on a parallel version of DTW.

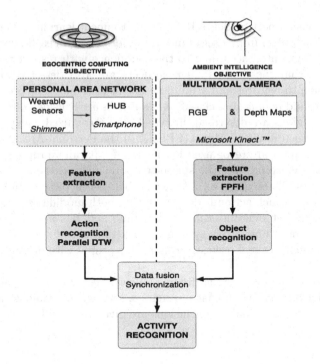

Fig. 1. Block diagram of the overall system.

Egocentric Features. Sensors used in the system have nine degrees of freedom (accelerometer, gyroscope and magnetometer with 3-axis each one). Selected features are raw inputs given from accelerometer and gyroscope and also their acceleration and angular speed energies. Magnetometer data is disregarded since we found it is not useful in gesture recognition.

Parallel Dynamic Time Warping. The problem of processing the information in real-time is solved using a variant of Dynamic Programming. The system has to process a large amount of input data (50 samples per second with 8 values and a time-stamp). In this sense, we use a parallel version of Dynamic Time Warping (DTW) [15] in order to speed up the activity recognition process. DTW algorithm provides a simple and fast way of aligning sequences. It computes the minimum path reconstruction of the input given a pattern. It is simply parallelized by taking each new sample as a possible beginning of a gesture. For each sample there will be a new thread processing it. We propose to store only the last two iterations and a reference to the beginning of the sample (i.e. the time-stamp). Then, when a pattern is accepted, we have the ending sample (with its time-stamp) and the reference we previously stored at the creation of the thread. This effectively identifies the gesture duration. Squared weighted Euclidean distance between the two eight-dimensional vectors is used as a composed metric.

The update rule is as follows,

$$M(i,j) = min(M(i-1,j), M(i,j-1), M(i-1,j-1)) + d(p_i, s_j),$$
$$d(p_i, s_i) = (p_i - s_i)^T (\omega^T I)(p_i - s_i),$$

where s_i is a sample vector, p_i is a pattern vector, M is the dynamic programming cost matrix, and I is the identity matrix. The weight vector ω is used to balance the importance of the accelerometer energy and gyroscope energy and to account for the different dynamic ranges of each feature. In order to speed up the performance of the algorithm, all threads computing a partial result higher than the acceptance threshold are finished at once.

2.2 Ambient Intelligence

The ambient module of the system is capable of detecting subjects that appear in the scene as well as to detect and recognize new objects appearing in the environment. This part is composed of 4 main submodules: environment modeling, user detection, object detection and recognition, and user-object interaction analysis.

Environment Modeling. A background subtraction strategy is applied. This allows to learn an adaptive model of the scene [6]. Given an initial set of multimodal frames, each one composed of a RGB image and a range image obtained from the Kinect™ device, a gaussian-distribution for each pixel is modeled.[1] Once the background has been modeled at pixel level, the apparition/removal of objects in the scene is detected whenever the pixels in a region show an absolute difference bigger than their learnt confidence values, i.e. greater than δ standard deviations.

User Detection. At each new frame, we perform user detection by segmentation using Random Forest approach on depth data. For this task, each cloud voxel captured from the scene is evaluated by a forest of trees trained on offsets of depth features. As a result, we obtain a user pseudo-probability for each point in the scene. From this, the user can be detected and an skeletal model as a spatial configuration of body limbs defined [4].

Object Recognition. Let the segmented image contain 1 at positions detected as foreground "objects" by the background subtraction ambient module. Each connected component of the segmented image which has not been classified as user is considered as a new object whenever its distance to the camera is smaller than the one obtained for the modeled background. In that case we compute a

---

[1] Note that due to the enrichment given by range data a single Gaussian model suffices for modeling background. Very small improvements have been observed using Gaussian Mixture Models in the studied environments.

normalized description of that particular 3D object view using the Fast Point Feature Histogram (FPFH) [7]. The FPFH is a point-wise 3D descriptor which describes the relative orientation of each point surface normal vector with respect to the average normal vector of the points in the k-neighborhood and encodes the information in a one-dimensional histogram. Those relative orientations are, basically, the roll, pitch, and yaw rotations discretized into 11 bins each one, summing up to a 33-bin descriptor. When a new object is detected in the scene, the descriptor for each point is computed to describe that particular object view. This description is compared to the data set of object descriptions using k-NN to classify the object.

2.3 Fusing Egocentric and Ambient Intelligence

Outputs from egocentric computing and ambient modules are synchronized in order combine the user, object, and activity recognition performed by both modalities. The synchronization is made using time-stamps adjusted with the NIST (National Institute of Standards) Internet time. Since an activity is defined as an action performed with a specific object, the intersection between object interaction and gesture recognition defines the activity the user is performing.

3   Results

In order to present the results, first, we describe the hardware, data and settings of the system in order to perform the experiments.

3.1   Hardware

The system is composed by a RGB-Depth camera (i.e. Microsoft® Kinect™) and a 9-degrees-of-freedom wearable Shimmer® device (Fig. 2), that includes triaxial accelerometer, gyroscope and magnetometer, Bluetooth® communication, up to 50 Hz sampling rate and the possibility to include additional modules (ECG, GPS, etc.). Also we used a Samsung GT-I9250 smartphone to stream the data to the PC running the server on a quad-core processor equipped with 8GB of RAM memory.

3.2   Data

We defined a dataset containing 13 different multi-modal synchronized recordings in 4 different types of scenes, with a length between 30 and 60 s. As a case of study for people with dementia, we include a scenario where the patient is taking his/her medication (a pill from a pillbox). This enables the evaluation of different objects, users and interactions in several scenarios.

In this case, the scenario has a table with three objects on it: a pillbox, a glass of water and a box. The camera is pointing towards the pillbox and the glass so the user will appear from one of the sides (see Fig. 2(a)).

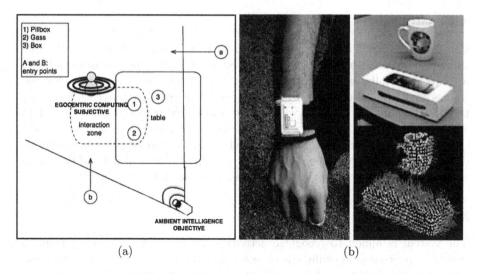

Fig. 2. (a) Setup of the ubiquitous system in a test scenario (b) On the left Shimmer® device tied to the arm of the user. On the right, the point clouds with the estimated surface normals of the objects (used in the FPFH object descriptor computation).

We have defined four possible scenarios for the validation: An unknown subject leaves the pillbox and a glass on the table. The patient appears in the scene. Then,

1. He takes the pillbox standing sideways from the camera, puts the pill on the mouth, leaves the pillbox on the table, drinks water, and leaves the room.
2. He takes the pillbox facing the camera, turns so that the camera cannot see what is doing, puts the pill on the mouth, leaves the pillbox on the table, drinks a bit of water, and leaves the room.
3. He takes the pillbox facing the camera, turns so that the camera cannot see what is doing, puts the pill on the mouth, leaves the pillbox on the table, drinks a bit of water, and leaves the room with the glass on the hand.
4. He takes the pillbox facing the camera, puts the pill on the mouth, leaves the pillbox on the table, drinks a bit of water, and leaves the room.

Although there are more than one scenario, the system only recognizes a single gesture.[2] Two important constraints of these recordings are: first, the camera has to see the patient taking the pillbox and leaving it to be able to detect the interaction. And second, the pill has to be taken by using the dominant arm, i.e. the one wearing the sensor. All objects, users and actions have been manually annotated in order to validate the performance of the system. The scenarios were performed by a single user without dementia.

---

[2] The extension to a small set of gestures of interest can be easily achieved without a significant loss in performance [11].

3.3   Settings

Parallel Dynamic Time Warping threshold is set using leave-one-out cross-validation to value 5.710^7, weights w used to compute the distance in DTW are experimentally set to 0.8 for raw data and 0.2 for the energy measures. 400 frames are used to learn a background model. Pixels segmented in each frame are those with change in its value with respect to the learnt model greater $\delta = 1.15$ standard deviations. A set of objects of interest is defined offline. Once a new object appears in the scene, it is classified as the nearest object minimizing he distance among the corresponding FPFH descriptors (see Fig. 2(b)) if this value is lower than 0.5 (otherwise it will be considered an unknown object).

3.4   Validation

The system is aimed to recognize activities. Recognizing an activity involves two main processes, knowing the object which the user is interacting with and also understanding the gesture performed during this interaction. In order to validate the system, we considered an overlapping measure, the Jaccard Index ($J = \frac{A \cap B}{A \cup B} = \frac{TP}{TP+FP+FN}$).
The different parts of the system work as follows:

- The interaction begins with the change of depth in the region of the object of interest. That is, moving the object out of its bounding box of the static position. The detection may take some frames from the beginning of the interaction (pick up event).
- In order to detect a new object, it has to be dropped and remain static for a few frames. This delay constraint in the object detection is set for the sake of robustness.
- Gesture detection is performed when the hand of the user starts a motion towards the mouth and returns to the first position.[3]

Figure 3 shows the average Jaccard index split in the four different scenarios, compared to the average system performance. Three bar sets show the individual performance measurements using only ambient data, egocentric data and the fusion of both. Observe that ambient analysis (multi-modal vision) achieve lower results. That is because in the vision system the taking-a-pill action is defined by the pick-up and release events but the action could be shorter if the users keep holding the object in his/her hands. This decreases the accuracy of the recognition since the detection could be done before the action begins or could exceed the action end. Additionally wearable sensor data usually performs well but not as well as the fused version. Notice the performance difference between scenarios 1–4 and 2–3 in the accuracy of the sensor. This happens when the subject is turning, that increases the length of the gesture of taking the pill. The fusion of the system maintains its robustness thanks to the help of the camera

---

[3] Notice that the drinking action is not detected because the system is sensitive to the hand orientation.

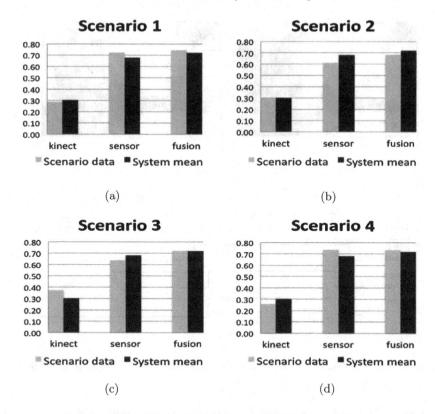

Fig. 3. Mean Jaccard value for each of the scenarios.

in the elimination of these outliers. Figure 4(a), (b) shows a real example of the system working. The interface of the application shows the depth map, the detected user, the detected objects in the scene, and detects the action of taking the medication (red square in (b)) shown by the inertial sensor features on the bottom of the images. Figure 4(c) shows an example of how fusion decreases the number of false positives. The patient touching his mouth is similar to taking a pill, thus it would be recognized by the sensor but not by the camera; in the fused data it would be a negative response.

Next, we include a discussion about the performance and reliability of the proposed system performing subject and object detection, object identification, and interaction and gesture recognition.

- The **subject detection** in RGB-D data by means of Random Forest (RF) has become a standard approach, being also exploited in commercial entertainment systems as Microsoft® XBOX360™. This technique provides very accurate results, and in our case it is reinforced by the usage of a background subtraction technique, since those regions not subtracted from the background are not considered people even if the RF detector gives a false positive, and thus discarded. From this, we cope with almost all the possible subject

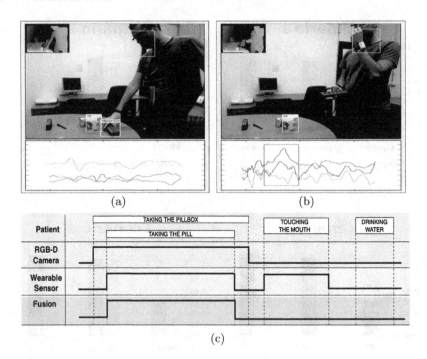

(a)   (b)

(c)

Fig. 4. (a) and (b) show object and gesture recognition. (c) Is an example of the fusion.

detection false positives except for the ones in non-human segmented foreground which actually never occur since we deal with small objects that differ from the random depth features learnt by the algorithm. Although the method can still give very few false negatives, it turns out to be very fast and accurate approach.

- The **object detection** performance is highly dependent on the capability of the system to accurately model the environment. Even though the infrared sensor of the Kinect[TM] device provides quite noisy measures, its price makes it a very good device to be considered in a wide range of applications. Because of these hardware limitations, the depth measures in a given pixel, even with the device fixed, range considerably. However, the variance in a certain pixel follows the same distribution throughout time, which we assumed to be normal. Then, to correctly model the pixels' variability a considerably large set of frames (400) is needed. Due to the noisy acquisition, very small objects can not be correctly modeled. Despite this, we have seen the system is able to detect objects with high precision if their area greater than 150 pixels in the dense depth map.

- The **object recognition** part presents the typical difficulties when performing in noisy and low-resolution images, as the case of dense depth images. One of the main limitations of the presented system is scalability, i.e. considerably increasing the number of objects to discriminate. In many applications, such as the presented one, we are interested in being able to classify among a small set of objects of interest. In this scenario, the method has shown a high classification accuracy performance.

– From the point of view of ambient intelligence, the **subject interaction with objects** task depends on the goodness of both the background subtraction and the subject detection. Since both of them perform correctly, and the interaction is straightforward to detect, the results are also accurate and precise.

– In the egocentric computing area, the **action recognition** is very accurate and delimits the gesture with high precision. However, in a future work we plan to include additional interaction and gesture recognition categories in order to analyze the scalability of generalization capability of both multi-modal vision and egocentric features.

4   Conclusion

We proposed to fuse egocentric and ambient features to define an integrated ubiquitous system to model daily actions of people with dementia. From the point of view of ambient intelligence, we learned a pixel-based Gaussian distribution of the background. Foreground segmentation is used to detect and recognize both user and objects based on 3D descriptor and statistical classifiers. From the egocentric point of view, we used a set of movement features computed from wearable sensors to test a parallelized Dynamic Time Warping gesture recognition method. We showed that fusing ambient and egocentric modalities provides a fast and robust system with high application potential, capable of recognizing different everyday activities involving different objects under the natural conditions of indoor scenes.

Acknowledgments. This work has been partly supported by RECERCAIXA 2011 Ref. REMEDI and TIN2009-14404-C02.

References

1. Stone, E.E., Skubic, M.: Evaluation of an inexpensive depth camera for passive in-home fall risk, assessment. In: Pervasive Computing Technologies for Healthcare (PervasiveHealth), pp. 71–11 (2011)
2. Zhang, C., Tian, Y., Capezuti, E.: Privacy preserving automatic fall detection for elderly using RGBD cameras. In: Miesenberger, K., Karshmer, A., Penaz, P., Zagler, W. (eds.) ICCHP 2012, Part I. LNCS, vol. 7382, pp. 625–633. Springer, Heidelberg (2012)
3. Banerjee, T., Keller, J., Skubic, J., Stone, E. E.: Day or nigh activity recognition from video using fuzzy clustering, techniques. IEEE Transactions Fuzzy Systems, pp. 1–1 (2013)
4. Shotton, J., Fitzgibbon, A., Cook, M., et al.: Real-time human pose recognition in parts from single depth images. In: CVPR, pp. 1297–1304 (2011)

5. Escalera, S.: Articulated motion and deformable objects 2012. In: Human Behavior Analysis From Depth Maps, pp. 282–292 (2012)
6. Clapés, A., Escaler, M., Reyes, S.: Multi-modal user identification and object recognition surveillance, system. Pattern Recogn. Lett. **34**(7), 799–808 (2013)
7. Rusu, R.B., Blodow, N., Beetz, M.: Fast point feature histograms (FPFH) for 3D registration. In: The IEEE International Conference on Robotics and Automation (ICRA), Kobe, Japan (2009)
8. Felzenszwalb, P.F., McAllester, D.A., Ramanan, D.: A discriminatively trained, multiscale, deformable part model. In: CVPR, pp. 1–8 (2008)
9. Ermes, M., Pärkkä, J., Mäntyjärvi, J., Korhonen, I.: Detection of daily activities and sports with wearable sensors in controlled and uncontrolled conditions. TITB **12**(1), 20–26 (2008)
10. Ouchi, K., Suzuki, T., Doi, M.: A wearable healthcare support system using user's context. In: Distributed Computing Systems, pp. 791–792 (2002)
11. Lichtenauer, J., Hendriks, E., Reinders, M.: Sign language recognition by combining statistical DTW and independent classification. IEEE Trans. Pattern Anal. Mach. Intell. **30**(11), 2040–2046 (2008)
12. Jiang, S., Cao, Y., Iyengar, S. et al.: CareNet: An integrated wireless sensor networking environment for remote healthcare. In: Body Area Networks, pp. 9:1–9:3, (2010)
13. Vintsyuk, T.K.: Speech discrimination by dynamic programming. Kibernetika **4**, 81–88 (1968)
14. Ming Hsiao, K., West, G., Venkatesh, S., Kumar, M.: Online context recognition in multisensor systems using dynamic time warping. In: ISNIPC, pp. 283–288 (2005)
15. Sakoe, H., Chiba, S.: Dynamic programming algorithm optimization for spoken word recognition. IEEE Trans. Acoust. Speech Signal Process. **26**(1), 43–49 (1978)
16. Pansiot, J., Stoyanov, D., et al.: Ambient and wearable sensor fusion for activity recognition in healthcare monitoring systems. In: 4th International Workshop on Wearable and Implantable Body Sensor Networks, pp. 208–212 (2007)
17. Stiefmeier, T., Ogris, G., Junker, H., Lukowicz, P., Troster, G.,: Combining motion sensors and ultrasonic hands tracking for continuous activity recognition in a maintenance, scenario. In: Wearable Computers, 2006 10th IEEE International Symposium, pp. 97–104, (2006)
18. You, S., Neumann, U.: Fusion of vision and gyro tracking for robust augmented reality registration. In: Virtual Reality, pp. 71–78, (2001)
19. Zhu, C., Sheng, W.: Motion- and location-based online human daily activity recognition. Perv. Mob. Comput. **7**, 256–269 (2011)

Transportation Planning Based on GSM Traces: A Case Study on Ivory Coast

Mirco Nanni[1]($\boxtimes$), Roberto Trasarti[1], Barbara Furletti[1], Lorenzo Gabrielli[1],
Peter Van Der Mede[2], Joost De Bruijn[2], Erik De Romph[2], and Gerard Bruil[2]

[1] KDD Lab, Isti CNR, Pisa, Italy
{mirco.nanni,roberto.trasarti,barbara.furletti,
lorenzo.gabrielli}@isti.cnr.it
[2] Goudappel Groep, Deventer, The Netherlands
{pvdmede,jdbruijn,edromph,gbruil}@goudappel.nl

Abstract. In this work we present an analysis process that exploits mobile phone transaction (trajectory) data to infer a transport demand model for the territory under monitoring. In particular, long-term analysis of individual call traces are performed to reconstruct systematic movements, and to infer an origin-destination matrix. We will show a case study on Ivory Coast, with emphasis on its major urbanization Abidjan. The case study includes the exploitation of the inferred mobility demand model in the construction of a transport model that projects the demand onto the transportation network (obtained from open data), and thus allows an understanding of current and future infrastructure requirements of the country.

1 Introduction

Population growth, massive urbanization and, particularly, the extensive increase of car use in the last century have led to serious spatial, transport, infrastructural and environmental problems in almost all urbanized areas. As a consequence, since the 1960s urban and transport planning methodologies were developed to forecast future traffic volumes and the expected use of infrastructure and facilities. The purpose of such forecasts is evident: infrastructure and urban planning provide keys to the mitigation and preclusion of transport and environmental problems. Today, urban and transport planning are major tasks of all public authorities.

The availability of spatial data on demographics, labor and land use has until now been a prerequisite for establishing an Origin and Destination matrix (OD matrix) for a transport model. It is a time consuming activity to obtain the necessary data in many developed countries. Moreover, in most developing countries the overall availability of data is very limited and, therefore, the use of transport models has never been a promising option for such countries. In this work we explore the possibility of deriving a proper OD matrix from mobile phone data, by using publicly available (free) transport network data and standard transportation modeling software, in order to build a basic transport demand model.

J. Nin and D. Villatoro (Eds.): CitiSens 2013, LNAI 8313, pp. 15–25, 2014.
DOI: 10.1007/978-3-319-04178-0_2, © Springer International Publishing Switzerland 2014

In this way, we also provide evidence that also for countries or cities where many data seem to be lacking, now transport demand models can be created. This will on the one hand allow national and local authorities to have a far better understanding of transportation needs and challenges, and will help funding agencies and investors to better assess their potential risks and benefits.

In the next sections we will describe step by step the process we propose to derive transport demand models from phone data, and use it to build a (as far as we know) first transport demand model for Ivory Coast and its major urbanization Abidjan.

2   Background

This work tries to combine data mining of GSM traces with transportation modeling methodologies to gain insights into mobility in a monitored area, to allow what-if analysis through simulation or modeling.

GSM data have already been used to describe mobility in several studies, essentially based on the fact that a sequence of geo-referenced calls of users constitutes approximate trajectories of their movements. The key limitations of GSM data are that locations are only approximations and that sampling rate may be low and erratic. Works like [1] try to overcome these issues by working at a large geographical scale and/or under specific conditions (in that case, users where tourists in a large area). In the present work we follow a different approach, and try to exploit the relatively long temporal extension of a dataset to infer more reliable movement information. In particular, an approach similar to [2] (translated from GPS to GSM data) [6] and [9] is adopted, where we try to extract regular movements that repeat consistently in time, which therefore are less likely to be artifacts of the data sampling procedure, and use them to measure systematic mobility in the area (details are provided in next sections). Also, concepts like most favored location, which are exploited in this work, have already been applied in the scientific studies, e.g. [3], but mainly for simple distributions of a population or the recognition of specific activities, such as working, being at home, or leisure.

Macroscopic transport modeling methodology is well established [7,8]. This methodology is mainly implemented through commercial and academic software tools (e.g. OmniTRANS, Visum, Cube, Emme/2, TransCAD), and readily available. These tools can be used only by professionals with accurate knowledge of traffic theory and transport modeling experience. Macroscopic modeling has been used widely by governments and engineering firms to predict future transport network problems and for infrastructure planning. The current paper does not address network or transport planning as such. Its main purpose is to use data mining of GSM traces as an input for transport models, thus integrating these traces as widely and readily available sources of information into the transport planning realm.

3   Introduction of the Case Study

The data used in this work is composed of anonymized Call Detail Records (CDR) of mobile phone calls and SMS exchanges between five million of Oranges customers in Ivory Coast (corresponding to around one quarter of the national population) between December 1, 2011 and April 28, 2012. The data was made available by Orange in the context of the D4D (Data for Development) data challenge [10]. The data contains 10 samples taken in different time windows, each covering 50,000 individuals, corresponding to around 1 % of the population of customers. The IDs of individuals are changed from sample to sample and it is not known whether the sub-populations described in the different samples overlap, making it impossible to link data among different samples. The data provided contains for each observation the coordinates of the antenna serving the user during a communication, in other words the device is operating in an area covered by that antenna.

In general the coverage of an antenna is influenced by several factors: strength of the signal, the height of the pole, the orientation, the weather, the nearby buildings, etc. Since the provided data does not contain this information and it is not easy to retrieve it from external sources, we apply a well-known methodology in order to estimate the coverage of the antennas using only their spatial location. The method is called as centroid Voronoi tessellation and assumes that the space is partitioned into separate areas, each defined as the set of locations that are closer to our antenna than any other one. The partitioning of the space obtained will be used in all the following analysis.

4   Systematic Traffic Analysis and Transport Modeling

The basic events we are interested to spot in the data are systematic trips. Following the approach in [2], we define systematic trips as routine movements that users perform (almost) every day at (approximately) the same hours. By combining together the systematic movements of each individual in our population, we can obtain an estimated OD matrix that describes the expected flow of people between pairs of spatial locations as in [4].

Since the current GSM data are not detailed enough to detect whether a user stopped at a location or initiated a trip within an input sequence, we tackled the problem through a two-step procedure: first, we identified locations that are significant for the mobility of the individual, also called attractors; second, we identified movements between significant locations that occur with a high frequency, which are later aggregated across the whole population to fill in an OD matrix. The first step is performed according to the standard approach, also illustrated in [3]: the location where the largest number of calls took place is identified and labeled as L1 (most frequent location). Then, the second most frequent location is identified and labeled as L2. It seems likely that in most cases L1 corresponds to the home location and L2 to work or any other main activity of the individual, or vice versa. The second step is performed over the sequences of L1 and L2 that appear in the traces of each single user. We checked the frequency of movements L1 L2 and L2 L1 within specified time slots. Each movement identifies a trip, and if its frequency is high enough, we assume it to be a systematic trip that the user performs during a typical day.

4.1   Detecting Users Attractors

The available data provide the information of the zone from which a phone call is started. Thanks to the large amount of data provided by the telephone operator it is possible to use the spatio-temporal footprint left by the users for the purpose of monitoring their movements in the territory.

Several studies [5] assert that most people spend most of their time at a few locations, and the most important ones may be labelled as *home* and *work*. In this section we will explain the methodology used for the extraction of such important locations, which we will call L1 (most important one) and L2 (second most important one) using the frequency of calls made by users. The location L1 relates to the antenna from which the user made the greatest number of phone calls. For both technical and infrastructure reasons due to the load balancing of the antenna, it may happen that the serving antenna for different calls made at the same place, may be different, even though such antennas are usually close to each other. To mitigate this effect, we have redefined L1 as a bigger area that also includes the adjacent cells. In Fig. 1, the most frequent location is represented by the central pink cell, but according to the method just described, we define as L1 the bigger area that includes the adjacent blue cells.

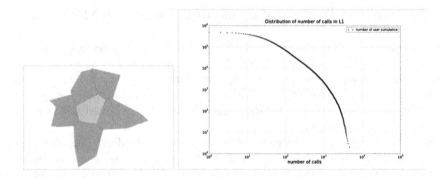

Fig. 1. Example of L1 detection (left), distribution of number of calls in L1 (right)

It is necessary to point out that, for the most of the users, the call frequency associated to L1 is quite low , thus rising issues of statistical significance and reliability of such a location (see Fig. 1). If we consider as a minimum frequency threshold for L1 of one call per day (therefore 15 calls in 15 days) from the area, only for 20 % of users (100 K) the associated L1 would result meaningful. The rest of this work assumes to use such subset of locations.

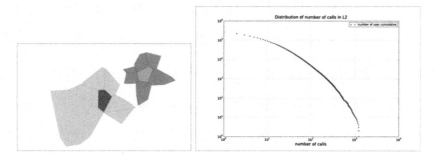

Fig. 2. Example of L1 and L2 detection (left), distribution of number of calls in the second most frequent location (right)

The L2 is defined as the area that ranked second in terms of call frequency, excluding all areas already absorbed into L1. Figure 2 Shows a visual example of L1 and L2. While in the example L1 and L2 are quite distant (for instance, these individuals might come to the city to work), in some cases they might be adjacent, which happens quite often in city centers where antennas are more dense. Also in this case it is necessary to consider a minimum support of phone calls to identify the significance of the places identified based on the distribution shown in Fig. 2(right). The result of this analysis will be used in the next step for the study of the systematic movements between preferred locations.

4.2 Detecting Systematic Movements

The focus of this analysis is the detection of systematic movements considering two separated time frames: a morning time frame, and an afternoon time frame, in which the users usually move, respectively, from home to work and from work to home. The first step is to identify the movements performed by individuals from L1 to L2 (L1 → L2) and from L2 to L1 (L2 → L1). It is important to notice that we are looking for movement between these two areas even if they are not contiguous, i.e. other areas were traversed between them, as shown in Fig. 3 (A is distinct from L1 and L2).

The second step consists in selecting only the systematic movements, which is done by applying two different constraints: (i) request a minimum number of movements between the pair; and (ii) request a minimum value for the *lift* measure [11] of the pattern L1 → L2, which we define as:

$$LIFT(L1, L2) = \frac{P(L1 \wedge L2)}{P(L1) \cdot P(L2)} \tag{1}$$

where $P(Li)$ ($i \in \{1, 2\}$) represents the frequency of Li, expressed as fraction of days where it appears at least once, and $P(L1 \wedge L2)$ represents the frequency of L1 and L2 appearing together. Lift measures the correlation between L1 and L2, resulting high if they appear together often w.r.t. the frequency of L1 and L2

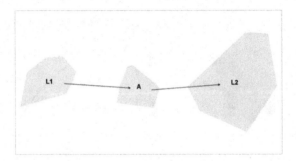

Fig. 3. Example of flow from L1 to L2

taken separately. The main purpose is to normalize the frequency of L1 → L2 w.r.t. the frequency of calls of the user, since otherwise the candidate movements of frequent callers would be excessively favoured in the selection. The threshold on the number of movements is usually adopted in literature to exclude extreme cases where the lift (or other correlation or relative frequency measures) is not significant. More important is the LIFT measure, in fact to select. In our case, after a preliminary exploration we chose to select only pairs that appeared at least 3 times. The threshold for the lift measure was chosen based on the study of its distribution, selecting the value where the slope of the cumulative distribution begins a sudden drop, corresponding to 0.7.

4.3   Systematic OD Matrix

As previously mentioned, the final goal of our analysis is the synthesis of O/D matrices that summarize the expected traffic flows between spatial regions. Our O/D matrices will focus systematic mobility, which represents the core (though not the only) part of traffic. In Fig. 4 some examples of intra-city traffic are shown, the first one from one origin to several different destinations, the second one from several origins to a single destination.

5   Application to the Case Study

In this section we summarize the process and results of inferring a transport model for Ivory Coast by applying the systematic mobility demand model extracted through the methodology illustrated in the previous section.

Network. We used data of the OpenStreetMap (OSM) road networks for Ivory Coast and Abidjan as a base for modeling. Although of great detail,in these network data many links are not, or not properly, connected. For route calculation this evidently is problematic. We therefore used an algorithm in a Geographic Information System (GIS) to find unconnected or badly connected roads and connected them properly or at least logically. Furthermore

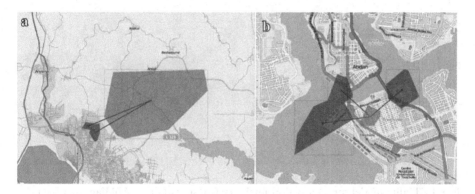

Fig. 4. Examples of traffic flows taken from one of the O/D matrices produced: (a) flows from the outskirts of Abidjian to the city; (b) from a single area to other districts.

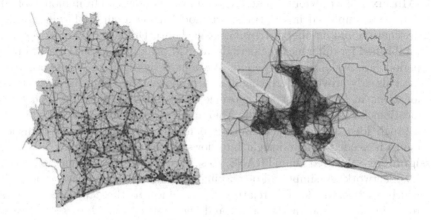

Fig. 5. Mobile phone movements in Ivory Coast and Abidjan.

we programmed an algorithm to identify small island-networks. They were either connected to the main network, typically we added a few ferries to connect real islands, or erased because they seemed illogical or unimportant. The resulting digital road network was imported in OmniTRANS, the software for transport modeling. Network link attributes, such as speed, which are necessary to calculate the shortest route between an origin and destination, were derived from link type information which is available from the OSM-network. The next step was to connect the antennas, e.g. the data carriers, to the network. In anticipation of this step we did not remove all the small roads from the network, which is often done in transport models. By keeping them we could simply connect each antenna to the nearest road. Of course this gives an overload on that particular minor road, but this quickly flattens out as all trips divert in different directions and converge on the major roads.

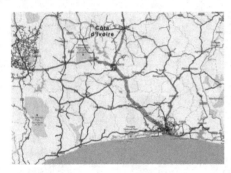

Fig. 6. Traffic model for 24 h period for Ivory Coast (left) and Abidjan area (right)

OD Matrix. The previously described OD-matrices derived from mobile phone data were imported in the transport model using a simple format (origin, destination and number of trips between them). Different OD-matrices were established for different time periods: an AM peak period (5–11), a PM peak period (15–21) and a 24h-period. Without assignment to a network these data already provide interesting images of movements. Figure 5 shows tower locations in Ivory Coast and movements of mobile phones between tower locations for Ivory Coast and the Abidjan area. These figures already provide a rough idea of the road network and major flows. However, a transport model is needed to gain insight into flows on the road network.

Assignment. We used OmniTRANS V6 software to assign OD-matrices to the road network. A simple all-or-nothing assignment technique was used by which all trips on an OD-relation are assigned to the calculated shortest route in time between the origin and the destination. More sophisticated assignment techniques are available (though requiring more data, like the road capacity), which take into account that different routes may be chosen because of congestion, but we felt the use of more sophisticated assignment technique was beyond the scope of this research.

Results and Interpretation of the Transport Model. Figure 6 shows assignments of the OD-matrices based on the GSM data for a typical 24 hour period for Ivory Coast and the Abidjan area. All major, national and urban transportation corridors are immediately visible from these plots. Also, the comparison between the linear movements between cell towers in Fig. 5 and the assigned movements in Fig. 6 provides a clear impression of the added value of assigning the movements to the road network. For purposes of readability we have decreased the level of detail in the figures in this paper. The original model plots allow much more detailed analysis of volumes on road links, in both directions of roads. Figure 7 shows traffic assignment on the network for the morning peak period in the greater (left) and central Abidjan area (right). As can be seen from all assignments the absolute flows, or traffic volumes on the network are very low and do not represent real traffic volumes on the network, since they are based on a selection of the sample provided for this study. Still, as a relative measure all figures show roads with more and less dense traffic. To make the transport model suitable for identifying and exploring current or future transport problems in Ivory Coast, an accurate assessment of absolute traffic flows on relevant parts of the network is necessary. In the following discussion we will deal with what must be done to overcome the limitations of the current model.

Discussion. It is highly rewarding that we were able to create a transport model for an area where we, as researchers, have never been, and for which data were solely obtained from the internet and from a completely new source (GSM call traces). However, the transport model is not yet finished, and this is mainly due to a number of remaining limitations in the data which were available. The good news is, that all these limitations can be solved, but the effort to do this varies.

First, to make an estimate of actual and validated traffic flows on the network, the current model values should be augmented or weighed by a factor. This factor will depend for instance on the market penetration of Orange operated cellphones, cellphone ownership and cellphone usage in Ivory Coast etc. As these levels may differ throughout the country, the weighing-method should take into account local differences. Techniques to do this are already available since also in transport modeling using conventional data, OD matrices are calibrated from traffic counts. A set of reliable traffic counts in the network should therefore suffice to establish augmented OD matrices to describe not just the traffic based on mobile phones traces, but the total traffic flows. A second, more serious limitation in the present model is that it does not make a distinction between the different transport modalities. To achieve such a distinction filters and algorithms must be developed to detect and estimate different modes of travel from data. The data available now would hardly allow a distinction of modes that is based on travel speeds. However, for modeling purposes, some basic statistics on modal split can be obtained from the same traffic counts as are needed to augment the traffic present in the model and could highly increase the quality of the model.

If we could overcome the above mentioned problems, and it seems absolutely feasible that this can be done, then we will have a model that can be used both for an accurate assessment of the current traffic situation in CI and Abidjan and for forecasting purposes. For forecasting purposes one needs to implement future planned projects in the model and specify the expected general growth in mobility for all modes. Once we have an adequately augmented mode specific OD matrix, the model immediately allows other calculations and forecasts for environmental impact analysis, such as greenhouse gasses, NOx and PM10 emissions. Of course, for these purposes additional statistics on the Ivory Coasts vehicle park must be incorporated.

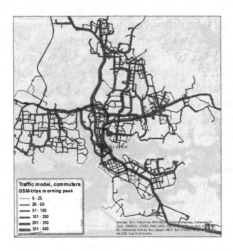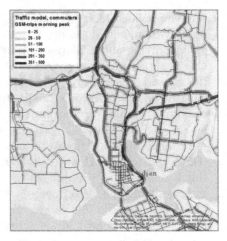

Fig. 7. Traffic model for morning peak period for greater Abidjan (left) and Abidjan Centre Area (right).

6   Conclusions

The present study shows that even with limited data from GSM traces, as in the case study considered, it is possible to derive valid information on systematic mobility behavior of people between frequently visited locations for areas that lack information on mobility. From these mobile phone data, origin-destination tables can be created for a chosen geographical area, which can then be used as an input for a transport model of the area. The fact that this can be done, overcomes one of the serious hurdles that until now have impeded the use of transport models in many developmental countries: lack of data. It is therefore absolutely worthwhile to take this proof of concept one step further, and create and validate a transport model that can indeed be used by public authorities, engineering firms, investors etc. in developmental countries. In this paper we discussed the most important steps which need to be taken.

References

1. Olteanu, A.-M., Trasarti, R., Couronn, T., Giannotti, F., Nanni, M., Smoreda, Z., Ziemlicki, C.: GSM data analysis for tourism application. In: ISSDQ (2011)
2. Trasarti, R., Pinelli, F., Nanni, M., Giannotti, F.: Mining mobility user profiles for car pooling. In: ACM KDD (2011)
3. Csji, B.C., et al.: Exploring the mobility of mobile phone users. Report on arXiv: 1211.6014 [physics.soc-ph]
4. Giannotti, F., et al.: Unveiling the complexity of human mobility by querying and mining massive trajectory data. VLDB J. (Special issue on Data Management for Mobile Services) **20**, 695–719 (2011)
5. Csjia, B.C., et al.: Exploring the mobility of mobile phone users. Physica A : Stat. Mech. Appl. J. **392**(6), 1459–1473 (2013)

6. Furletti, B., Gabrielli, L., Renso, C., Rinzivillo, S.: Identifying users profiles from mobile calls habits. In: UrbComp (2012)
7. Hensher, D.A., Kenneth, J.: Handbook of Transport Modelling. Pergamon, Amsterdam (2000)
8. Ortúzar, J., Willumsen, L.G.: Modelling Transport, 4th edn. Wiley, West Sussex (2011)
9. Calabrese, F., Di Lorenzo, G., Liu, L., Ratti, C.: Estimating origin-destination flows using mobile phone data. Pervasive Comput. 4, 36–44 (2011)
10. D4D Challenge. http://www.d4d.orange.com/
11. Tan, P.-N., Steinbach, M., Kumar, V.: Introduction to Data Mining. Addison Wesley, Boston (2006). ISBN 0-321-32136-7

From Tweets to Semantic Trajectories: Mining Anomalous Urban Mobility Patterns

Lorenzo Gabrielli[1], Salvatore Rinzivillo[1],
Francesco Ronzano[2], and Daniel Villatoro[2]([✉])

[1] Knowledge Discovery and Data Mining Laboratory, ISTI-CNR, Pisa, Italy
{salvatore.rinzivillo,lorenzo.gabrielli}@isti.cnr.it
[2] Mobility and Energy Group, Barcelona Digital Technology Centre,
Barcelona, Spain
{fronzano,dvillatoro}@isti.cnr.it

Abstract. This paper proposes and experiments new techniques to detect urban mobility patterns and anomalies by analyzing trajectories mined from publicly available geo-positioned social media traces left by the citizens (namely Twitter). By collecting a large set of geo-located tweets characterizing a specific urban area over time, we semantically enrich the available tweets with information about its author – i.e. a resident or a tourist – and the purpose of the movement – i.e. the activity performed in each place.

We exploit mobility data mining techniques together with social network analysis methods to aggregate similar trajectories thus pointing out hot spots of activities and flows of people together with their variations over time. We apply and validate the proposed trajectory mining approaches to a large set of trajectories built from the geo-positioned tweets gathered in Barcelona during the Mobile World Congress 2012 (MWC2012), one of the greatest events that affected the city in 2012.

Keywords: Trajectory analysis · Social media · Urban mobility · Geographic data mining

1 Introduction: Mining Urban Mobility

The digital breadcrumbs produced by individuals while using modern ICT services provide us with the opportunities of tracking personal behaviors with an unprecedented granularity. Thanks to the modern social networks, like Facebook, Twitter, Foursquare, it is possible to observe each individual from different points of view, considering for instance her mobility, social status and relationships, preferences [1].

In this paper, we address the problem of reconstructing collective mobility patterns by the retrieval and analysis of publically available opportunistic data sources: in particular, we investigate approaches to sense the mobility of people by mining the collection of geo-located tweets crawled from a specific geographic area [2]. By properly aggregating geo-located tweets, we are able to reconstruct

J. Nin and D. Villatoro (Eds.): CitiSens 2013, LNAI 8313, pp. 26–35, 2014.
DOI: 10.1007/978-3-319-04178-0_3, © Springer International Publishing Switzerland 2014

the movement of users, i.e. the trajectory they followed to move across several locations [3]. Location-based social networking has attracted a lot of attention in the last years, since it provides an effective repository of crowd-sourced behavioral information [4,5]. The different approaches in the literature address the discovery of extraordinary events in near-real time by analyzing the content of the tweeted messages [6,7], or providing analytical methods to understand individual mobility [8].

We exploit the repertoire of mobility data mining techniques to analyze such movement data [9]. By accessing DBpedia, we characterize each trajectory as performed by a local person or a tourist. Moreover, the information describing these trajectories is also increased by pointing out the activity the user is going to do in her destination as well as by the activity the same user was performing before moving. To this purpose, we exploit Foursquare to associate a candidate venue and thus a specific venue category related to the activity performed by the same user, to both the origin and destination of each displacement. We rely on the geographic, temporal and semantic information related to trajectories to get a vision of user mobility from both the spatiotemporal and the semantic perspective thus deriving proper insights.

To demonstrate the validity of our approach we selected a set of tweets collected in three consecutive weeks in the metropolitan area of Barcelona, where the central week presents a burst in the number of tweets produced. To motivate this unusual behavior we build trajectories from the set of geo-located tweets and we analyze the temporal evolution of collective mobility within the city looking for regularities as well as for relevant anomalies. In particular, we manage to clearly mark and delimit geographically the presence of such a large event in the city and, by exploiting the semantic annotation of the activities characterizing each trip, we are able to better understand the type of the event we have discovered. In particular, the three weeks corresponds to the period around the Mobile World Congress 2012 (MWC2012), one of the biggest events occurred in 2012 in Barcelona.

The paper is organized as follows. In Sect. 2 we describe the MWC2012 dataset. Section 3 presents the main analytical approaches we adopt to mine mobility patterns and the main results emerging from their application to the MWC2012 dataset. Section 4 explains how we can get useful urban mobility insights exploiting the semantic characterization of trajectories and validates this approach by exploiting the MWC2012 dataset. Section 5 summarizes the analysis presented in this paper and sketches future work.

2   Extraction Twitter Trajectories

To demonstrate the analytical approach we are presenting, we collected around 183 k geolocated tweets generated by 11 k distinct users in the metropolitan area of Barcelona during a period of three weeks. Geo-located tweets concerning these three weeks have been gathered by means of bounding-box filtered Twitter Streaming API.[1]

Trajectories have been created by aggregating two or more geo-located tweets produced by the same user in locations more than 100 m far away one from the other, within a time interval lower than 75 min. The spatial threshold was determined by analyzing the distribution of spatial gaps among two consecutive tweets of the same user. The temporal threshold was chosen in relation of the maximal duration of a ticket of the public transportation system. In this way, a total of 9689 trajectories have been identified: 2991 during week 0, 4139 during week 1, and 2559 during week 2.

Fig. 1. Enriching geo-located tweets by associating Foursquare venue/category

Each one of these trajectories has been enriched with the following information (see Fig. 1):

- **Local/Tourist Flag**: by considering the location of the Twitter user associated to each trajectory, we classified the trajectory as related to a local inhabitant or a tourist. In particular, we performed such annotation with two distinct strategies. The first approach uses DBpedia[2] to retrieved the set of *Populated Places* (such as cities, towns or villages) in the Catalonia Region by means of SPARQL queries. The second approach, exploits the information associated with the account of each user to deterime his/her home location. After manual refinement and enrichment of this dataset, we built a classifier able to point out if a Twitter user is Catalan or not (thus a tourist) by analyzing the information contained in his Twitter user profile. In this way, over 9689 trajectories, we managed to classify as local or tourist related 7236 trajectories (83,3 %). The details of the outcomes of this process are summarized in Table 1. To estimate biases of locals versus tourists, we checked the observed number of trajectories per user in the three week for the two categories. Figure 2 shows that the distribution of the number of trajectories per user are comparable for the two categories of users.

---

[1] Twitter Streaming API: https://dev.twitter.com/docs/streaming-apis/parameters# locations

[2] DBpedia: http://dbpedia.org/

Table 1. Number of trajectories by Week, considering All trajectories, Tourist trajectories and Local trajectories

Trajectories	All	Tourist	Local
Week 0	2991	525	1453
Week 1	4139	1437	1749
Week 2	2559	537	1535

– **Foursquare Venue Top Category of Origin and Destination**: by exploiting Foursquare API[3] we retrieved the most-likely Foursquare venue that can be associated to the GPS position of the origin and the destination of each trajectory (see Fig. 3). In particular, given a GPS location (for instance the one of a tweet), the Foursquare API provides an ordered list of locations that could be related to it: we consider the first/most-likely location present in such list, relying on the outcomes of the research on venue proximity ranking supporting the same Foursquare API [10]. In this way, for almost the complete set of trajectories included in our dataset (99.83 %), we managed to associate both origins and destinations to a Foursquare venue. Since Foursquare venues are classified by means of a hierarchy of categories, it is possible to retrieve for each venue the root of this hierarchy, referred to as Foursquare Top Category. In particular, Foursquare venues classification includes 9 Top Categories: *Nightlife Spot, Travel & Transport, Outdoors & Recreation, Shop & Service, College & University, Food, Arts & Entertainment, Residence, Professional & Other Places*.

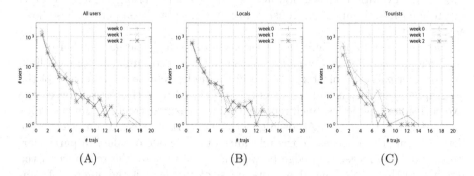

Fig. 2. Distribution of the number of trajectories per user in the three weeks for (A) all users, (B) locals, and (C) tourists

---

[3] Foursquare API: https://developer.foursquare.com/

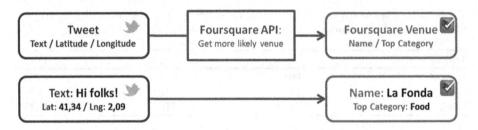

Fig. 3. Enriching geo-located tweets (origin and destination of trajectories) by associating Foursquare venue/category

3   Hot Spot Identification by Geographic OD Matrix Analysis

The identification of relevant zones associated with urban anomalies like large events is performed by means of the analysis of variations in regular patterns of movement. We exploit the trajectories built from the geo-located tweets (see Sect. 2) to analyze flows within the city of Barcelona in successive time periods in order to automatically identify variations within a specific time period. For the identification of flows across time, we choose to adopt OD matrix analysis in order to deterministically extract geo-referenced moves from a set of fixed geographic regions. In particular, we consider the districts included in the administrative territory of the city. For each pair of such regions we estimated the observed number of trajectories starting in the first one and ending in the second one. Since the spatial tessellation is fixed a priori, it is straightforward to observe the evolution of each flow across different time periods. Thus, our strategy to identify relevant flows consists in monitoring and comparing these flows across the time.

However, we do not limit the observation only to the actual number of trajectories for each flow but we address also the structural influence of each link. We consider each OD matrix as a graph: the nodes represent the regions of the geographic tessellation and the links are associated with the flows between the two corresponding nodes. Such representation is easily extracted from the information of the OD matrix. According to this network representation, we can use network statistics to measure the relevance of each node or edge. In particular, we adopted the measure of edge betweenness [11] to state the relevance of the link within the graph. The edge betweenness of an edge e is the number of minimum paths traversing that specific edge, where the set of minimum paths is computed by determining the path for each pair of nodes in the graph. Edge betweenness provides a more robust measure than the count of trajectories of each flow, since it takes into account the topological structure of the network and it is capable of handle local spikes within a single link.

In Fig. 4 we show the distribution of the edge betweenness considering all the edges for the three consecutive weeks of the dataset. On the x axis it is reported the list of all the links sorted by the edge weight decreasing. Each chart represents the statistics for two distinct populations: tourists or visitors, and residents. There are significant increments in edge betweenness in week 1 if compared to week 0 and week 2. These differences are even more evident when considering tourist flows. In fact, it can be noted how the betweenness for week 1 is much higher than the other two weeks. In the chart, we have highlighted the origin and destination of the most relevant flows. In particular, the flows that show a sensible increment are related to two regions: Eixample and Sants-Montjuic. We will focus our analysis on these two flows to better understand the reasons of this increment in the week.

We consider two classes of users: locals and tourists, as described in Sect. 2, and we distinguish among flows related to both classes. From Fig. 4, we can notice that the increment related to the regions Eixample and Sants-Montjuic is more evident for the movements of tourists (Fig. 4 - A) during week 1, than locals (Fig. 4 - B). Thus, the cause of this augmented mobility is influenced primarily by persons visiting the city, probably for a specific event happening in that district.

Moreover, if we consider the semantics associated to each visited location according to Foursquare classification, we can notice that the people entering the Eixample district come to perform a preeminent activity, namely an activity associated with Professional venues. Figure 5 - A shows the distribution of the activities performed by people entering Saint-Montjuic district. These individuals came mainly to venues associated with *Professional & other* category, i.e. to perform something associated with their job. Once leaving this area, the activities associated with their destinations are mainly associated with *Food* and *Shopping* (Fig. 5 - B). This evidence reinforces our hypothesis that these are people arrived in the city for a specific event linked to their job. By looking into the event lists of that week in the city we found an event that is compatible with the hypotheses found so far: the annual Mobile Week Conference that took place in the Sants-Montjuic area.

4   The Purpose of Trajectories: Analysis of the Semantic OD Matrix

Thanks to the association of the origin and the destination of each Twitter-mined trajectory to the most-likely Foursquare venue, it is possible to characterize urban displacements by considering their semantics. In particular, we perform our analyses by taking into account the Foursquare Top Category of the venues associated to the origin and the destination of each trajectory: as a consequence we are able to build the Semantic Origin Destination matrix of the 9 Foursquare Top Categories. By means of this matrix, we manage to characterize both the relevance of the activities carried out at the beginning and at the end of each trajectory together with the importance of these activities with respect to the urban context.

We exploit both a classical tabular representation as well as a chord diagram to visualize the Semantic OD matrix of the 9 Foursquare Top Categories. In

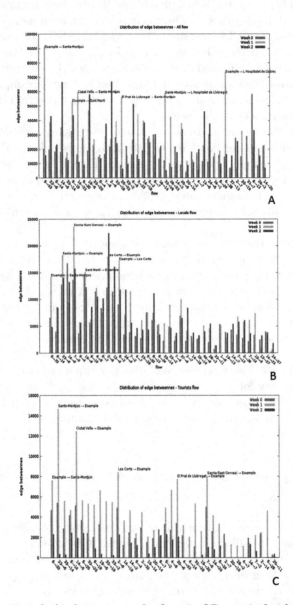

Fig. 4. Distribution of edge betweenness for flows in OD matrix for the three consecutive weeks of the MWC2012 dataset (week 0: week before week 1: week of MWC2012 week 2: week after) and three distinct populations: residents (B), tourists (C), and all (A)

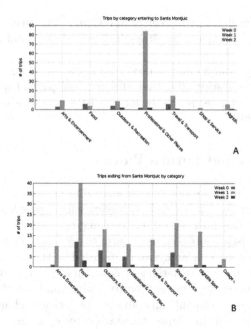

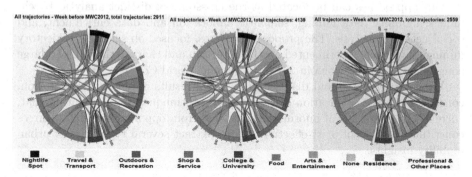

Fig. 5. Distribution of activities per location: (A) activities of people arriving to Sants Montjuic; (B) activities of people leaving Sants-Montjuic

Nightlife Spot   Travel & Transport   Outdoors & Recreation   Shop & Service   College & University   Food   Arts & Entertainment   None Residence   Professional & Other Places

Fig. 6. Chord diagrams of Semantic Origin Destination Matrix of Twitter trajectories during the week of MWC2012, the week before and the week after

particular, Fig. 6 shows the chord diagram of the Semantic OD matrix concerning all the trajectories related to the area of Barcelona during the week of MWC2012, the week before and the week after the event.

An interactive Web visualization of the Semantic OD matrix of the 9 Foursquare Top Categories including the division of trajectories across tourist and local ones can be accessed at: http://penggalian.org/semanticTrajectories/. From Fig. 5 we can notice how the relevance of trajectories related to *Professional & Other Places* is considerable during the week before MWC2012 and reaches its maximum during the week of MWC2012. The week after MWC2012, the relevance of these trajectories drops in favor of *Food* and *Shop & Services* ones.

This fact clearly points out the professional nature of the event that takes place in the city, suggesting also that the greatest part of attendees reached Barcelona the week before, probably for touristic activities or to prepare the location for MWC2012.

5   Conclusions and Future Works

In this paper we address the problem of analyzing human mobility in order to identify relevant patterns, flows and anomalies within people displacements in urban scenarios. To this purpose we retrieve, aggregate and semantically enrich data from opportunistic sources so as to form trajectories: in particular we consider geo-located tweets. Our analyses are tailored to the automated identification and characterization of variations of flows and urban activities in order to point out relevant mobility patterns, identify outliers and link them to events. To this purpose, we consider semantically enriched trajectories characterizing urban mobility during a defined time lapse: these trajectories are mined by building of OD matrices, by analyzing the edge betweenness of the resulting graph and by proposing and exploiting a semantic version of the same OD matrix.

Our approaches can be located at the crossroad of distinct analytical techniques: mobility data mining, semantic annotation and data enrichment, and social network analysis. The proposed analyses, focused on semantic trajectory mining, have been experimented for the detection and characterization of a large congress held in Barcelona in 2012, the Mobile World Congress.

We have presented and discussed the initial results of our experimentation, consequent to the application of the proposed techniques, demonstrating that, currently, the analysis of information gathered from opportunistic data sources constitutes a relevant, cost-effective way to extract several typologies of urban mobility insights.

Acknowledgements. This work has been completed with the support of ACC1Ó, the Catalan Agency to promote applied research and innovation; and by the Spanish Centre for Development of Industrial Technology under the INNPRONTA program, project IPT-20111006, "CIUDAD2020". The work was also partially supported by the EU project DATASIM N. 270833.

References

1. Giannotti, F., Pedreschi, D., Pentland, A., Lukowicz, P., Kossmann, D., Crowley, J., Helbing, D.: A planetary nervous system for social mining and collective awareness. Eur. Phys. J. Special Topics **214**(1), 49–75 (2012)

2. Villatoro, D., Serna, J., Rodríguez, V., Torrent-Moreno, M.: The tweetbeat of the city: microblogging used for discovering behavioural patterns during the MWC2012. In: Nin, J., Villatoro, D. (eds.) CitiSens 2012. LNCS (LNAI), vol. 7685, pp. 43–56. Springer, Heidelberg (2013)

3. Java, A., Song, X., Finin, T., Tseng, B.: Why we twitter: an analysis of a microblogging community. In: Zhang, H., Spiliopoulou, M., Mobasher, B., Lee Giles, C., McCallum, A., Nasraoui, O., Srivastava, J., Yen, J. (eds.) WebKDD/SNA-KDD 2007. LNCS, vol. 5439, pp. 118–138. Springer, Heidelberg (2009)

4. Lee, R., Sumiya, K.: Measuring geographical regularities of crowd behaviors for twitter-based geo-social event detection. In: Proceedings of the 2nd ACM SIGSPATIAL International Workshop on Location Based Social Networks, LBSN '10, pp. 1–10. ACM, New York (2010)

5. Lee, R., Wakamiya, S., Sumiya, K.: Discovery of unusual regional social activities using geo-tagged microblogs. World Wide Web **14**(4), 321–349 (2011)

6. Sakaki, T., Okazaki, M., Matsuo, Y.: Earthquake shakes twitter users: real-time event detection by social sensors. In: Proceedings of the 19th International Conference on World Wide Web, WWW '10, pp. 851–860. ACM, New York (2010)

7. Chae, J., Thom, D., Bosch, H., Jang, Y., Maciejewski, R., Ebert, D., Ertl, T.: Spatiotemporal social media analytics for abnormal event detection and examination using seasonal-trend decomposition. In: 2012 IEEE Conference on Visual Analytics Science and Technology (VAST), pp. 143–152 (2012)

8. Fuchs, G., Andrienko, N., Andrienko, G.: Extracting personal behavioral patterns from geo-referenced tweets. In: AGILE 2013 (2013)

9. Giannotti, F., Nanni, M., Pedreschi, D., Pinelli, F., Renso, C., Rinzivillo, S., Trasarti, R.: Unveiling the complexity of human mobility by querying and mining massive trajectory data. VLDB J. Int. J. Very Large Data Bases **20**(5), 695–719 (2011)

10. Shaw, B., Shea, J., Sinha, S., Hogue, A.: Learning to rank for spatiotemporal search. In: Proceedings of the sixth ACM International Conference on Web Search and Data Mining, pp. 717–726. ACM (2013)

11. Brandes, U.: A faster algorithm for betweenness centrality. J. Math. Sociol. **25**(2), 163–177 (2001)

Incentivising Crowdsourced Parking Solutions

Andrew Koster[1]([✉]), Fernando Koch[2], and Ana L.C. Bazzan[1]

[1] Institute of Informatics, Federal University of Rio Grande do Sul (UFRGS),
Rio Grande do Sul, Brazil
{akoster,bazzan}@inf.ufrgs.br
[2] SAMSUNG Research Institute Brazil, Campinas, Brazil
fernando.koch@samsung.me

Abstract. The problem of finding parking slots imposes both societal and infrastructural issues in modern cities. It is a daily hurdle that affects millions of people, but existing approaches fail to solve this conundrum. Thus, there is an urgent demand for reputable, motivated, and replicable solutions that can be used by cities of any size. We are proposing an experiment to analyse the interplay between incentive mechanisms, user participation, and the truthfulness of reports. For that, we are developing the "wePark application" based on concepts of crowd sourcing and social regulation. As a differential, we are examining alternative methods to motivate adoption, such as reciprocity, reputation, altruism, and money. In this paper, we analyse the requirements of the solution, propose a development test bed, and an experimental environment for this study.

1 Introduction

The problem of finding free parking imposes both (a) social issues, due to the annoyance caused to citizens, and (b) traffic infrastructure impact. For instance, [10] reports that around 40 % of traffic in New York is generated by cars searching for parking spaces. Different approaches have been proposed such as the SFpark [6] in San Francisco, which provides centrally controlled "parking sensors" to identify free parking in real-time. However, even with the large budget of this project, only a small part of San Francisco is covered. Meanwhile the advent of commonly available smartphones, with GPS and online capability, allows for citizens to fulfill the role of *sensor*, should the proper motivation exist for them to provide such information. Google released an Android application called *Open Spot* [3], which allow users to report and find free parking spaces. However, this proposal failed because people did not correctly report enough free parking spaces [7]. This scenario reinforces the demand for reputable, motivated, and replicable solutions that can be used by cities anywhere.

However, there is a lack of understanding on how to engineer a working solution applying participatory sensing [4] approaches for this problem. We hypothesise that it will require a balance between crowd sourcing, reputation models, and incentive mechanisms. We propose an experiment based on a mobile application (to collect data) being use in a controllable environment, and analytic

J. Nin and D. Villatoro (Eds.): CitiSens 2013, LNAI 8313, pp. 36–43, 2014.
DOI: 10.1007/978-3-319-04178-0_4, © Springer International Publishing Switzerland 2014

models to analyse social behaviour and reputation parameters. We want to better understand the influence of motivation mechanisms, social behaviour, and social interactions in this problem domain. In specific, we aim to address two research questions: (i) how to incentivise citizen participation in such projects, and (ii) how to model and understand individual behaviour whilst utilising these systems?

To that end, we are developing the "wePark application" for reporting and finding parking spaces in a specific area. The solution works based on crowd sourcing concepts and allows for participants to provide feedback on others' reports, providing an interface for social networking. As a differential, we are promoting alternative methods to incentivise adoption, such as reciprocity, reputation, altruism and money. Moreover, we are focusing on the development of reputation models, trying to identify and filter misleading reports, which will help to promote credibility and usage.

We intend to validate this approach in a "Living Campus Experiment" along with the UFGRS campus. That is, we want to create the infrastructure of Smart City technologies to transform the campus in a living lab, integrating mobile computing resources, open services, and advanced Analytic Models similar to the proposal in [5], previously conceptualised by IBM Research. This setup provides a unique resource for joint research, allowing the development of meaningful field tests and fast turn around.

This paper is structure as follows. Section 2 provides an overview of related research and the prior art. Section 3 present the "wePark Application" as a tool to conduce our research. Section 4 concludes with the expected results and analysis.

2 Background

There are a number of research projects in the domain of crowdsourcing to find a suitable incentive model in order to attract participation, such as the ones described in [1,9]. Nonetheless, it is an ongoing research topic and very much domain dependent. Yan et al. [11] propose a market-based approach for traffic, where people who are leaving their parking space can sell information about its location to people in need of a parking space. However, this must be done in advance, and drivers must know when they are leaving, or arriving, to sell or buy a parking space, respectively.

A more general approach, that is also more similar to the approach taken by Google's *Open Spot* and similar commercial Apps, is proposed by Chen et al. [2]: participation itself is the incentive mechanism. However, they propose a number of improvements over commercial apps in order to make the user's experience better, and thus more likely to use the service. Nevertheless, their evaluation is simulation-based, so it is unclear whether this is truly sufficient.

Tokarchuk et al. [9] have categorised the motivations of people participating in crowdsourcing activities as follows:

- Reciprocity and expectancy
- Reputation
- Altruism
- Self-esteem and learning
- Fun and personal enjoyment
- Implicit promise of future monetary rewards
- Money

Other references provide slightly different lists, but in general agree. Not all these motivational methods are available to different applications, and even when available, may be more, or less, effective. It may also not always be obvious what incentive works. For instance, the *Old Weather project* [8] found that they could motivate people by giving them a story to read, thus motivating participants with personal enjoyment: something they were not expecting in a serious Citizen Science project.

One way of motivating people to report free parking spaces is to improve the functioning of the app, thus lowering the "cost" of reporting while simultaneously increasing the utility of the provided information. This is similar to the proposed in Chen et al. [2]. Alternatively, Yan et al. [11] proposes to provide a monetary incentive. Nevertheless, these have not been tested in an real-world experiment leaving a gap for experimentation.

In our project, we hypothesise that the main motivating factors for reporting parking spaces are: *reciprocity*, *reputation*, *altruism*, and *money*. Altruism and reciprocity seem straightforward: people are reporting parking spaces out of sympathy for others, or by installing the app they intend to use it to search for parking spaces when they are in need, and hope others will be reporting them (reciprocating). This also seems to be the main reason *Open Spot* and similar programs failed: these motivating factors were not enough. Nor was Google's reputation incentive, by awarding so-called Karma points for reporting free parking spaces.

Therefore, we propose an experiment to put this to the test in a controlled environment, in order to discover what can incentivise participation.

3   Proposal

In order to conduce this experiment, we are proposing to research and develop the *wePark application* and analytic models to understand social behaviour and reputation parameters. This environment will develop upon the concepts of Citizen Participation, Social Networking and Community Engagement.

The architecture of this framework is depicted in Fig. 1. The application will run as a crowd sourcing solution, allowing for end-users to: (i) easily report free parking spaces, and; (ii) search for a parking space nearby or at a distant location (related to his programmed trip). In addition, the back-end solution will implement the algorithms that provide the intelligence to the system, such as reputation ranking, filtering of parking places reported by, and to, different users, clustering of reports to avoid reporting the same spot multiple times, and

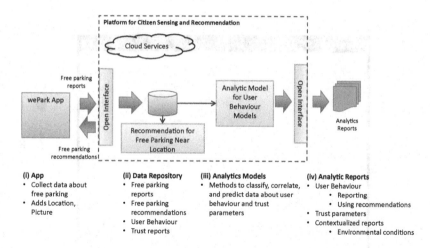

Fig. 1. Architecture for the wePark experimental framework

other possible processes for improving the functioning of the system. We will test how this intelligence can be used to ensure the social balance, by detecting and eliminating misleading information and erratic behaviour.

We intend to use the "wePark application" to gather data on users' behaviour and improve the application incrementally. This data set will provide insight into how similar crowd sourcing applications could be used for dissemination of other traffic information, such as the location of traffic jams. The advantage of this approach is that there is very little known about user behaviour in such situations. We start from the principle that for some crowd sourcing applications the intrinsic incentives are enough to be successful, while others require external incentives to be added, such as credit systems. Similarly, whether users' reports are truthful depends on many different factors, which are largely unknown in the traffic environment.

The app will be integrated with route planner software, and the initial iteration will have the following functionality:

- One-click reporting: if the user clicks the report button, it will report a free parking space at the user's current GPS coordinates, and at the current time.
- Automatic parking space display: any reported parking spaces that are near the route's destination will be displayed on the map. These will be colour coded according to the time they were reported. The map can be browsed through as well.

Figure 2 depicts the screenshot of a prototype. The features being introduced in this application provide improved functionality over other solution, making it easier to report parking spaces, as well as incorporating them into the navigation software. We purposefully do not propose to use any other incentives, such as Google's *Karma points*, or monetary incentives for reporting a parking space.

Fig. 2. Prototype wePark application

We also do not intend to punish malicious users in any way, because we are particularly interested in how the app is used.

3.1 Trial Setup

The user trial will follow students and university employees on the UFRGS Campus do Vale in Porto Alegre, Brazil. The main advantage of deploying this in the Living Campus context is because it is a restricted environment in which a controlled user experiment can be conducted. After monitoring users' participation throughout a month we will ask them about their experience with the app, and compare this with the collected data. This will serve as input for a next iteration, in which incentive mechanisms can be added, we can sanction malicious users or improve the app in other ways.

While we do not know for sure how many users we will include in the trial, the aim is to reach as many commuters as possible in order to have a sufficient group for accurately reporting parking spaces on the campus. We will initially distribute it among CS students and professors, but make it available for download and do some advertising of the app on the campus to disseminate information about it.

We realize that a university campus has a different demographic to the city as a whole and any conclusions we draw will have to be confirmed in a larger population. Nevertheless, as an initial field trial environment, the advantages of a small area with a large number of drivers who are easily contacted outweigh this disadvantage.

3.2   Collected Data

The data being collected consists of usage data, as well as answers to a questionnaire. The usage data will record how often, when, and where participants report parking spaces, or open the app to search for a parking space. We will also collect a user's route when using the app, and can match this to either successfully finding a parking space, or driving past spaces that have been reported as free. The questionnaire will ask for the user's subjective opinion on their use of the app, with questions such as: "are you pleased with your usage of the report function?", "what are the main reasons for not reporting a parking space?" and "did you ever falsely report a parking space? If so, why?" to discover what might better motivate users to truthfully report parking spaces. Similar questions will serve to discover if users are pleased with the experience of finding a parking space with the app, and what can be done to make it better.

The usage data will then be coupled with questionnaire answers to discover usage profiles, which we will classify, roughly, according to what motivates the participants. We can then improve the app with further incentives to reinforce these motivations. The improved app will be tested in a second iteration of the experiment, thus allowing us to test specific hypotheses about the motivations of users and how to incentivise participation in a collaborative software.

4   Conclusion

In this paper we analysed the requirements of a solution for the free parking problem by studying the interplay between incentive mechanisms, user participation and the truthfulness of reports. We are developing wePark, a proof-of-concept solution for reporting and finding parking spaces, which provides the data collecting capabilities in our platform. As a differential, we are promoting alternative methods to incentivise adoption, such as reciprocity, reputation, altruism and money. The empirical experiment that we propose is designed to answer the following two questions: (i) how to incentivise citizen participation in such projects, and (ii) how to model and understand individual behaviour whilst utilising these systems. We presented a prototype proposal and discussed some design decisions, such as provided advanced end-user experience, not creating any monetary incentives, and implementing on-line surveys to model user profiles. We will be applying this solution in a "Living Campus" experiment at UFRGS, collect data, and apply analytic models to analyse how user behaviour and social interaction impacts the utilisation of this solution. Our objective is to engineer a working solution applying participatory sensing approaches, and make it work based on the balance between crowd sourcing, reputation models, and incentive mechanisms.

Acknowledgements. Andrew Koster is supported by CAPES (PNPD) and Ana Bazzan is partially supported by CNPq. Dr Koch was a Research Scientist with IBM Research Brazil when this study was first designed, thus we applied concepts provided by IBM Research.

References

1. Brabham, D.C.: Moving the crowd at istockphoto: the composition of the crowd and motivations for participation in a crowdsourcing application. First Monday May 2008 (2008)
2. Chen, X., Santos-Neto, E., Ripeanu, M.: Crowd-Based smart parking: a case study for mobile crowdsourcing. In: Borcea, C., Bellavista, P., Giannelli, C., Magedanz, T., Schreiner, F. (eds.) Mobilware 2012. LNICST, vol. 65, pp. 16–30. Springer, Heidelberg (2013)
3. Hildenbrand, J.: Google releases open spot for android – find and share parking (10 July 2010). http://www.androidcentral.com/googl-releases-open-spot-android-find-and-share-parking. Accessed June 24 2013
4. Boulos, K.M., Resch, B., Crowley, D., Breslin, J., Sohn, G., Burtner, R., Pike, W., Jezierski, E., Chuang, K.Y.: Crowdsourcing, citizen sensing and sensor web technologies for public and environmental health surveillance and crisis management: trends, OGC standards and application examples. Int. J. Health Geographics 10(1), 1–29 (2011)
5. Koch, F., Cardonha, C., Gentil, J.M., Borger, S.: A platform for citizen sensing in sentient cities. In: Nin, J., Villatoro, D. (eds.) CitiSens 2012. LNCS, vol. 7685, pp. 57–66. Springer, Heidelberg (2013)
6. SFpark. http://sfpark.org. Accessed June 25 2013
7. Tedeschi, B.: Cutting through the bother of city parking. New York Times (9 March 2011). http://www.nytimes.com/2011/03/10/technology/personaltech/10smart.html. Accessed June 24 2013
8. Thompson, K.: All hands on deck. Sci. Am. **306**(2), 56–59 (2012)
9. Tokarchuk, O., Cuel, R., Zamarian, M.: Analyzing crowd labor and designing incentives for humans in the loop. IEEE Internet Comput. **16**(5), 45–51 (2012)

10. Transportation Alternatives: No vacancy: Park slope's parking problem and how to fix it. http://www.transalt.org/files/newsroom/reports/novacancy.pdf. Accessed June 24 2013 (2007)
11. Yan, T., Hoh, B., Ganesan, D., Tracton, K., Toch, I., Lee, J.S.: Crowdpark: a crowdsourcing-based parking reservation system for mobile phones. Technical report UM-CS-2011-001, University of Massachusets (2011)

Transportation Networks

Crowdsensing Simulation Using ns-3

Cristian Tanas[✉] and Jordi Herrera-Joancomartí

Universitat Autònoma de Barcelona, Barcelona, Spain
ctanas@deic.uab.cat, jordi.herrera@uab.cat

Abstract. A crowdsensing network is a sensor network in which sensors are users that sense the environment and send the obtained data using, for instance, their smartphones. The performance of such sensor networks depends heavily on the mobility of the users and their willingness to collaborate. It is hard to obtain a stable set of users to evaluate such kinds of sensor networks and, for that reason, studies of crowdsensing networks are scarce. In this paper, we describe how the ns-3 network simulator can be used to simulate some crowdsensing networks with specific characteristics by using the mobility properties of network nodes together with the wireless interface in ad hoc network mode. We model the identification of network nodes with users of the crowdsensing network and we define how to simulate user sensing capabilities. Finally, we present a simulation example for a specific crowdsensing network where users report incidents in the public rail transport.

1 Introduction

Mobile sensing has become one of the most appealing areas in sensor network research over the last decade as the penetration of mobile devices in our daily life activities reaches unprecedented levels [13].

Indeed, smartphones, besides being sophisticated computing platforms, come along with a complete set of sensing tools (positioning -GPS-, motion -accelerometer-, image -camera-, audio -microphone-, among others), empowering these devices to sense and monitor their surrounding environment. Moreover, smartphone owners should not be ignored since they can also detect distinct properties of their surrounding environment and their human-readable descriptions could complement hardware sensor readings, leading to a new level of sensing, that is *smart sensing*. Relying on users to sense the environment and send the obtained data using their smartphones leads the way to a new generation of sensor network, commonly referred as *crowdsensing networks*, where users become *crowdsensors*.

Therefore, we can leverage the power of collective intelligence, as stated by Cardone et al. [9], together with the sensing capabilities of these smart devices in order to build more powerful and richer sensor networks. Although crowdsensing networks can help overcome many of the limitations of existing proposals in sensor networks, their performance strongly depends on the mobility of the users and their willingness

This work has been partially supported by the Spanish Government through project TIN2010-15764 N-KHRONOUS and the UAB grant PIF 472-01-1/E2010.

J. Nin and D. Villatoro (Eds.): CitiSens 2013, LNAI 8313, pp. 47–58, 2014.
DOI: 10.1007/978-3-319-04178-0_5, © Springer International Publishing Switzerland 2014

to cooperate. In other words, the number of users in a crowdsensing network and their mobility patterns determine unequivocally the area that can be sensed. Also, persuading users to participate in the sensing tasks of a crowdsensing network is highly challenging and, for that reason, only a few crowdsensing network application proposals can be found in the literature and all with an experimental nature. For instance, the MetroSense project [8] offers a network architecture for people-centric sensing that harnesses existing urban infrastructure and human mobility to opportunistically sense environmental data, with applications such as BikeNet [11] or SkiScape [10]. However, none of the applications provide any estimation of the number of users or sensor density required to assure their usefulness.

Simulation may seem a viable solution to analyse the performance of a crowdsensing network application before to launch a real deployment. There are many software-based simulation platforms available, including *ns-2* and *GloMoSim* which are the two most popular ones, that provide a simplified model of the real world and a complete environment to conduct extensive simulations in a wide range of possible scenarios. Even though there are many studies for wireless or mobile ad hoc networks (MANETs) simulation [18], there are no proposals to simulate people-centric sensor networks. Nevertheless, due to the similarities of crowdsensing networks with MANET scenarios, we can still use existing network simulators to analyse the behaviour and performance of specific crowdsensing network applications, and to study the impact that parameters such as sensor density or mobility patterns have on these new kind of environments.

In this paper, we present the *ns-3* network simulation platform focusing on its capabilities of modelling, simulating and analysing the performance of a wide range of crowdsensing networks. We harness the mobility and wireless communication properties of the simulator in order to simulate user sensing capabilities and bring a sense of crowdsensing applications into the simulated world. Furthermore, we introduce a new mobility model for users whose motion behaviour is constrained by the underlying infrastructure of the crowdsensing network and provide a simulation example of a real-world crowdsensing application.

The particular kind of sensing scenarios we bring under discussion in this paper are those where users sense their environment for events that are discrete in time and space, regardless of the events' semantics. That is to say, events occur at precise moments of time and at particular locations, and they can only be detected by the users in the nearby area. Then, the overall goal of the crowdsensing network is to detect and inform of all the events that take place in the area it is supposed to cover. Although this may seem a hard simplification, many smart sensing applications, focusing on the presence or absence of an event and allowing an a posteriori processing of the semantics, fall into this category.

This paper is organized as follows. In Sect. 2, we review some of the most popular network simulation tools. Section 3 presents the architecture of the *ns-3* network simulator and a procedure for crowdsensing network simulation with special emphasis on node mobility. In Sect. 4, we introduce a new mobility model for crowdsensors based on graph theory. A simulation example is provided in Sect. 5. Finally, Sect. 6 concludes the paper.

2   State of the Art

In order to provide an insight of the tools we can use in smart sensing-related research, and without claiming completeness, this section presents some of the most popular simulation platforms, both commercial and open-sourced, that can be used in order to simulate the behaviours of a crowdsensing network.

Most network simulators follow an open source philosophy in order to enable researchers to contribute with new models and protocols to the simulation platform. Two of the most popular network simulators in this category are NS-2 and GLO-MOSIM. The former is considered to be the *de facto* simulation tool for networking research, including wireless and ad hoc scenarios [3], as confirmed by Kurkowski *et alter* in [18], while the latter offers a discrete-event simulation platform targeting wireless network systems and suitable for large-scale wireless network simulations [27].

In the open source category we can also find **J-SIM** [1], which provides a distinct architecture built upon the notion of the autonomous component architecture (ACA) [25]. Its design principles mimic those of an integrated circuit where different components are assembled together and interactions between these components occur to simulate specific behaviours. However, if performance is a requirement, a better choice would be **SWANS** [4], which is a scalable wireless network simulator built on top of the JiST simulation engine [5]. A notable characteristic of this simulator is that it can run regular, unmodified Java network applications (web servers, multicast protocols, ...) over the simulated network.

Another open source network simulator that is worth mentioning is **GTNETS**, a discrete-event simulation tool targeting general networking research [21]. The architecture and design of the simulator match those of a real network protocol stack and hardware, and is implemented entirely in C++.

There are also simulation tools that only allow application level protocols in the ISO/OSI Reference Model to be simulated, assuming that an approximate emulation of the lower levels is available, including **DIANEMU** [17] and **JANE** [14, 15].

A more recent simulator is **THE ONE** [16], a Java-based network simulator designed to simulate specific Delay-Tolerant Network protocols and applications [12]. It provides a rich framework for simulating scenarios where node mobility is a critical characteristic.

As for commercial suites for network simulation, **QUALNET** [24], **OPNET** [22], and **OMNET++** [23] are worth mentioning. All of them provide a complete set of user-friendly tools for protocol programming, simulation configuration and data visualization. Although they are commercial solutions, they also support academia and research licences which are free of charge.

Despite the great number of available simulators, the majority of them fail to provide a good, in-depth documentation of their architecture and the procedure required to contribute new models or application protocols.

On the other hand, every crowdsensing application scenario has to be simulated under realistic assumptions, being the mobility of the sensors the most compelling one. In order to simulate that mobility we can use either traces or synthetic models. Traces correspond to motion behaviours observed in real-life systems, while synthetic models aim at providing a realistic behaviour of mobile users motion. Although traces are much more realistic, obtaining them is a challenging task. Therefore, the most used models are synthetic ones [7]. Typically, random-based models are used for performance assessment, such as *Random Walk* or *Random Waypoint* where mobile users are assumed to move towards new locations by randomly selecting a direction and speed, and additionally a pause time, in the case of the Random Waypoint model. Furthermore, the *Gauss-Markov* mobility model is often used in order to avoid the erratic motion behaviour presented by the above models since it uses a parameter that tunes the degree of randomness in the mobility pattern.

Most of the existing proposals in synthetic mobility models are already implemented in the majority of network simulators. Nevertheless, new models can be used by means of mobility trace files which contain the motion behaviour of all the mobile nodes in the simulation employing a specific format. A widely used mobility definition format is the *ns-2* mobility format, which provides a simple set of instructions that can be used to indicate the initial and future locations of mobile nodes, as well as their speed. There are several tools that can generate movement trace files using this format, including BonnMotion [6], SUMO [2] or TraNS [19].

3   ns-3 Overview

ns-3 is a completely renewed version of the discrete-event *ns-2* network simulator, making a huge step forward in organization and performance from its predecessor. Designed for both networking research and education, *ns-3* provides a new, well documented core implemented entirely in C++ adding new models, based on well-known abstractions, including a Wifi link type and several mobility models.

ns-3 is built as a library and its functionalities are divided into separate modules which may be statically or dynamically linked to a C++ main program that defines the simulation topology and conducts the actual simulation. Moreover, it provides Python wrappers to conduct simulations using the Python programming language. Existing models can be more or less heavily modified and even new models can be built from scratch using the C++ programming language in order to enable richer simulations.

The core architecture of *ns-3* emulates the real behaviour of complex network systems, including the ISO/OSI protocol stack. The key abstractions defined by the simulator correspond to the most commonly used concepts in networking, such as nodes, applications, transmission channels, etc. So, the basic unit of interaction throughout the simulation is the *Node* which encapsulates the behaviour of any computing device that connects to a network. Therefore, we can resemble the behaviour of web servers, data storage facilities or even individuals carrying a smartphone by referring to them as Nodes in the simulation. Nodes by themselves perform no actions and have to be assembled in a similar way a manufacturer customizes its devices. Figure 1 depicts the basic model of interaction in *ns-3*. As we can see, each Node runs Applications that define its behaviour, has its own protocol stack following the ISO/OSI model, and has peripheral network interface cards (NICs) installed to enable it to communicate with other Nodes in the same network. Additionally, Nodes may be static or mobile, in which case they have associated a mobility pattern that specifies the motion behaviour of the Node around the simulation area.

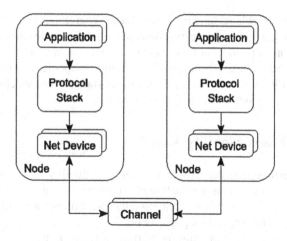

Fig. 1. Basic model of interaction in *ns-3*

3.1 Modelling a Crowdsensing Network

As we have mentioned before, a crowdsensing network is a sensor network where users sense their surrounding environment and send the sensed data using their smartphones. We focus on a specific kind of sensing, that of events that are discrete in time and space. We suppose that a crowdsensing node is able to detect the presence of an event if and only if he or she falls within a given range from the location of the event. Moreover, collective knowledge of the event can be derived from the observations of all the crowdsensing nodes in the area determined by the maximum range at which the event can be detected, considering that users can share information through their smartphones' wireless communication properties.

Given the above assumptions, we can map a user of the crowdsensing network in the real world to a *Node* in the *ns-3* architecture. Then, the simulation will consists of as many nodes as crowdsensors whose behaviour we want to analyse. Every Node in the simulation will have attached a specific *Application* that will emulate the sensing behaviour of the users of a particular crowdsensing network. Therefore, the *Application* will drive the simulation of the interactions between these users.

We can take advantage of the *Node's* wireless communication capabilities in order to enable communication between crowdsensors. In this way, the action of sensing an event and sending the collected information trough the smartphone will be mapped to a network packet sent via a wireless communication channel. Moreover, we can leverage the *ad hoc* mode[1] of a wireless NIC to ensure that the sent data will only reach those nodes that are in the nearby area. The meaning of a "nearby area" can be quantified by restricting the communication range of network Nodes. *ns-3* provides a wireless channel implementation that follows the physical layer model described in [20] and, typically, simulates a wireless channel with a propagation delay equal to the speed of light and a propagation loss based on a distance model, which defines the maximum range at which

---

[1] All devices are assumed to have equal status in the network and are free to communicate with any other ad hoc device in link range.

a network Node is capable of sending wireless data. Therefore, by restricting a node's maximum communication range we can simulate the maximum distance at which a user of the crowdsensing network could sense the occurrence of an event.

Furthermore, we can take advantage of the mobility properties of a network Node to simulate the mobility of users in a crowdsensing network. A more extensive discussion regarding mobility is provided in the following section.

4   A Graph-Based Mobility Model

Crowdsensing performance depends heavily on the mobility of its users since, for instance, users mobility determines the bounds of the area that the crowdsensing application can cover. Any events, whose location fall outside of this area will remain undetected for the users of the crowdsensing network.

We can benefit from the mobility properties of network nodes in order to simulate the mobility of users in the real world. *ns-3* provides extensive support for user mobility and mobility patterns can be applied to network nodes in different ways. On the one hand, implementation of several well-known mobility models are available for simulation. These include *Random Walk*, *Random Waypoint* or *Gauss-Markov*, among others. Nonetheless, new models can be built from scratch or by combining some of the existing models. On the other hand, mobility models can be specified using the *ns-2* mobility format. A separate mobility trace file must be provided at the configuration phase of the simulation, which the simulator parses in order to set the mobility pattern for each node in the simulation. Note that this method is only valid for those scenarios where the mobility of the crowd is not affected by its sensing activity. In other words, a node can not dynamically alter its destination based on the information sensed from the surrounding environment.

The most used mobility models in wireless or ad hoc network simulation are random-based (e.g. Random Walk, Random Waypoint, etc.). However, the motion behaviour of the users in a crowdsensing application highly differs from such models. Instead, the mobility of users is strongly influenced by their daily activities and by the infrastructure of the area in which they are located. Under these assumptions, we present a graph-based approach to model the mobility of crowdsensing network users, taking into account the time constraints of their daily life activities and those of the infrastructure underlying the sensor network.

We denote by $G = (V, E)$ the graph that represents the underlying infrastructure of a particular crowdsensing network. The set of vertices of the graph, $V = \{v_1, \cdots, v_n\}$, corresponds to the n locations that users may visit, while the set of edges, $E = \{e_{ab} = (v_a, v_b)$ for $v_a, v_b \in V\}$, map all the possible routes between these locations (e.g. streets or train connections). We also denote by $\omega(e_{ab})$ the weight of the edge connecting vertices v_a and v_b, which represents the relation between the locations mapped to those vertices (e.g. distance, time). Figure 2 depicts an example of such a graph modelling the Barcelona underground service.

Our mobility model is based on paths over the graph G. A path of the graph G is a possible route between two given vertices of the graph, v_{i_1} and v_{i_m}, without revisiting any vertex:

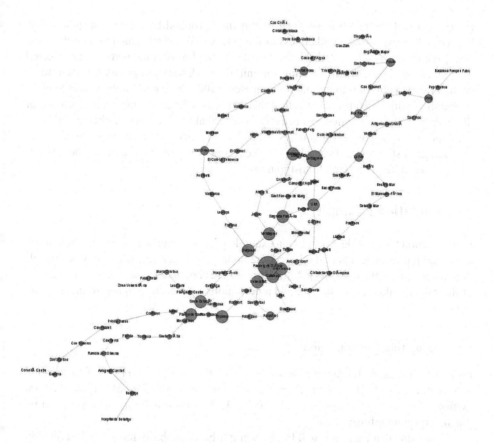

Fig. 2. Example of a graph modelling the underground railway system of the metropolitan area of Barcelona

$$P_{i_1 i_m} = \{(v_{i_1}, \cdots, v_{i_m}) \text{ where } v_{i_j} \in V,\ e_{i_j i_{j+1}} \in E\}$$

Since the path can be seen as an edge sequence, a straightforward definition of the cost of a path is the sum of the weights of its edges:

$$\Omega(P_{i_1 i_m}) = \sum_{e_{i_j, i_{j+1}} \in P_{i_1 i_m}} \omega(e_{i_j, i_{j+1}})$$

In order to determine the mobility of the users from vertex v_a to vertex v_b, assuming that more than one path between those nodes may exist in G, we take the one that minimizes $\Omega(P_{ab})$, that is the shortest path. Depending on the definition of the edge weight, the shortest path may minimize time, distance or even other variables such as economic cost.

Besides the path that a user follows to go from one node to another, a simulation environment should define which are the initial and final nodes that have to be assigned to each user of the crowdsensing network. In order to avoid a random selection, we select initial and final destination points for each user through different probabilities.

Every vertex of the graph is assigned two distinct probabilities: an *initial probability*, $\mathcal{P}_{ini}(v_a)$, that expresses the likeliness of the vertex to be selected as initial point, and a *final probability*, $\mathcal{P}_{end}(v_a)$, which determines the likeliness of the vertex to be selected as destination. Then, when selecting the initial and destination points for each node, each vertex is subjected to an independent Bernoulli trial, based on the corresponding probability, which determines whether the vertex is selected as the initial location of the user or as the destination respectively. Notice that the journeys of citizens in a city exhibits such a behaviour as vertices located in the city center are more likely to be selected as destination points, while vertices located in the outer areas of the city are more likely to be the origin of most journeys.

5   Simulation Example

In this section we provide a simulation example of a crowdsensing network where users sense and report incidents in the Barcelona underground service, while they are travelling towards their places of work. Simulations were conducted using the *ns-3.14* version of the *ns-3* simulator, in which we add a new application corresponding to this sensing scenario.

5.1   Simulation Configuration

First of all, we define the simulation area to match the area covered by the underground service of the city of Barcelona. Then, we perform several simulations varying the total number of users (crowdsensors), from 100 to 1200, whose mobility is constrained by the underground infrastructure.

We assume that each user will follow a graph-based mobility model as described in Sect. 4, using a *GraphML* description of the graph seen in Fig. 2. Such graph consists of $|V| = 116$ vertices corresponding to underground stations and $|E| = 262$ edges corresponding to the underground connections between the stations. The weight of each edge, $\omega(e_{ab})$, is defined as the distance (in meters) between the locations represented by vertices v_a and v_b respectively.

The mobility of each user is defined by the shortest path between a pair of stations, probabilistically sampled from the set of all station, following a Poisson sampling model. However, initially, each user will be placed at a distance of 300 m from his or her initial selected station. We use an initial probability distribution based on vertices' degree, assuming that vertices with higher degree are more likely to be selected as destination points as they represent an intersection point of several underground routes. Therefore, vertices having a degree greater or equal to 6 are assigned a final probability $\mathcal{P}_{end}(v_a) = 0.9$, while their initial probability $\mathcal{P}_{ini}(v_a) = 1 - 0.9$. In other words, if 3 or more underground routes intersect in one station, the vertex corresponding to this station will be selected as a destination point with probability 0.9. On the contrary, vertices having a degree less than 6 are assigned a final probability $\mathcal{P}_{end}(v_b) = 0.1$, while their initial probability is $\mathcal{P}_{ini}(v_b) = 1 - 0.1$. This behaviour tries to model a mobility scenario where people living in the outer zones of Barcelona travel by underground to their jobs located mostly in the center of the city.

In order to provide another degree of realism to the simulation, we assume that not all users begin their journeys at the same time, since the timetable restrictions may vary for each user. Therefore, we define an one hour interval for users to begin moving towards their initial assigned station. Then, the final simulation time is determined by the maximum instant of time a user reaches his or her destination.

Incidents, the events that have to be sensed, are generated every 100 s at a randomly-chosen vertex of the graph. Then, we examine all the users whose journeys include the selected vertex between 3 min prior and after the occurrence of the event. From the set of crowdsensors satisfying this requirements, we select one at random which will be responsible for generating a new incident report. Moreover, we demand other users in the area to confirm the generated incident in order to consider it valid. However, we establish a maximum communication range of 100 m. Consequently, only those users located at less than 100 m from the location of the incident will be able to confirm it.

As the users move around the simulation area, the number of crowdsensors available at a specific location will vary with time, defining the node density around the location of the event. Such parameter bounds the maximum number of observations that can be gathered from that event. Moreover, node density is of extreme importance in a crowdsensing network since a value below 1 would mean that most of the events generated would remain undetected.

5.2 Simulation Data Results

Next, we provide the results of our simulation analysis regarding the configuration described in the previous section. In order to avoid biased results due to the standard deviation of the random number distributions used during the simulation, we repeat each simulation 10 times and compute the final results as the average of the results obtained in each individual stage.

First of all, we want to account the total number of incidents that are detected by the users of the crowdsensing network. In Fig. 3 we present the number of detected incidents based on the total number of users of the crowdsensing network. As we can see, the percentage of detected incidents increases from 29 %, in a crowdsensing network formed by 100 users, to nearly 70 % in a crowdsensing network formed by 1200 users. It is worth to notice that even with 100 crowdsensors we can still detect nearly 30 % of the incidents, which is quite acceptable given the large area of the considered scenario. Moreover, since the final simulation time is defined by the maximum instant of time a user reaches his or her destination and the paths that users follow towards their destination may vary significantly in length, the actual number of active crowdsensors at the end of the simulation may fall dramatically.

Using our simulation scenario we can also evaluate the sensor density depending on the total number of sensors for our system. Table 1 illustrates the results, showing the total number of users required to reach a given average user density around the location of an incident. We can observe that the number of users needed grows in a linear way as we increase the required average user density. However, notice that in order to reach a sensor density of 6 around the location of an incident, we only need over 1000 users. That means when an incident is generated there are, on average, 6 users of the crowdsensing network at a distance less than 100 m from the location of the

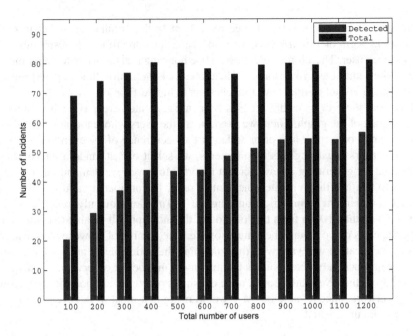

Fig. 3. Number of detected incidents by the users of the crowdsensing network

Table 1. Average node density by total number of users

Total users	100	**200**	300	400	600	800	**1000**	1200
Node density	1.42	**1.91**	2.45	3.15	4.12	5.15	**5.98**	7.01

incident. Therefore, if we request an incident to be confirmed by other users to consider it valid, the sensor density becomes a critical parameter and, in the case of an 1000 users crowdsensing network, we could reach a 5 confirmation threshold. In addition, if our scenario requires just an 1 confirmation threshold for any incident, we could reach that threshold with little over 200 crowdsensors. Although 1000 crowdsensors may seem a large number, we can compare it with the number of users travelling by underground within the metropolitan area of the city of Barcelona supplied by *TMB*, the company that provides the service. In its 2011 annual report [26], *TMB* estimates that nearly 389 million users travelled by underground during that year, which gives us around 1.07 million users travelling by underground in one day (on average). Therefore, 1000 users represent roughly 0.001 % of the estimated underground users on an average day.

6   Conclusion and Further Research

The wide spread and use of smartphones unfolds great potential to effectively map human-centric sensing task to end-user controlled smartphones. However, the performance and usefulness of such sensor networks depends heavily on the mobility characteristics of users and their willingness to cooperate. Moreover, it is hard to deploy a crowdsensing network in a real scenario and validating design principles for crowdsensing networks is a highly challenging task. In this paper, we have presented a software-based approach for crowdsensing network simulation, which allows researchers to analyse the performance of such sensor networks without having to deploy the sensing application on real devices. Furthermore, we propose a graph-based mobility model that takes into account the underlying infrastructure of a crowdsensing network to define mobility patterns for its users. We also provide a simulation example of a crowdsensing network application where users report incidents in the underground service of Barcelona. The data obtained from the simulation shows us that with little over 1000 users, we could detect nearly 70 % of the incidents that occur in the underground system. Nonetheless, predicting the mobility patterns of a crowdsensing network raises complex security and privacy-related issues that we intend to analyse and bring under discussion in further research.

References

1. J-Sim. https://sites.google.com/site/jsimofficial/. Accessed 5 June 2013
2. Simulation of Urban MObility (SUMO). http://sourceforge.net/apps/mediawiki/sumo/index.php?title=Main_Page. Accessed 27 May 2013
3. The Rice University Monarch Project: Mobile Networking Architectures. http://www.monarch.cs.rice.edu/. Accessed 27 May 2013
4. Barr, R.: An efficient, unifying approach to simulation using virtual machines. Ph.D. thesis, Cornell University, May 2004
5. Barr, R., Haas, Z.J., van Renesse, R.: Jist: an efficient approach to simulation using virtual machines. Softw.: Pract. Experience **35**(6), 539–576 (2005). http://dx.doi.org/10.1002/spe.647
6. University of Bonn: BonnMotion. A mobility scenario generation and analysis tool. http://net.cs.uni-bonn.de/wg/cs/applications/bonnmotion/. Accessed 27 May 2013
7. Camp, T., Boleng, J., Davies, V.: A survey of mobility models for ad hoc network research. Wireless Commun. Mob. Comput. **2**(5), 483–502 (2002)
8. Campbell, A., Eisenman, S., Lane, N., Miluzzo, E., Peterson, R.: People-centric urban sensing. In: Proceedings of the 2nd Annual International Workshop on Wireless Internet, WICON '06. ACM, New York (2006)
9. Cardone, G., Foschini, L., Bellavista, P., Corradi, A., Borcea, C., Talasila, M., Curtmola, R.: Fostering participaction in smart cities: a geo-social platform. IEEE Commun. Mag. **51**(6), 112–119 (2013)
10. Eisenman, S., Campbell, A.: SkiScape sensing. In: Proceedings of the 4th International Conference on Embedded Networked Sensor Systems, SenSys '06, pp. 401–402. ACM (2006)
11. Eisenman, S., Miluzzo, E., Lane, N., Peterson, R., Ahn, G.S., Campbell, A.: BikeNet: a mobile sensing system for cyclist experience mapping. ACM Trans. Sen. Netw. **6**(1), 6:1–6:39 (2010)
12. Fall, K.: A delay-tolerant network architecture for challenged internets. In: Proceedings of the 2003 Conference on Applications, Technologies, Architectures, and Protocols for Computer Communications, SIGCOMM '03, pp. 27–34. ACM, New York (2003), http://doi.acm.org/10.1145/863955.863960
13. Ganti, R.K., Ye, F., Lei, H.: Mobile: current state and future challenges. IEEE Commun. Mag. **49**(11), 32–39 (2011)

14. Gorgen, D., Frey, H., Hiedels, C.: Jane-the java ad hoc network development environment. In: 40th Annual Simulation Symposium, ANSS '07, pp. 163–176 (2007)
15. Görgen, D., Frey, H., Hiedels, C.: Jane-a simulation platform for ad hoc network applications. In: Demos of the 9th International Symposium on Modeling, Analysis and Simulation of Wireless and Mobile Systems (2006)
16. Keränen, A., Ott, J., Kärkkäinen, T.: The ONE simulator for DTN protocol evaluation. In: SIMUTools '09: Proceedings of the 2nd International Conference on Simulation Tools and Techniques. ICST, New York (2009)
17. Klein, M.: Dianemu - a java based generic simulation environment for distributed protocols. Technical report, Sixth International Workshop on Network-Based Information Systems (NBIS2003), in the framework of the 14th International Conference on Database and Expert Systems Applications 2003 (2003)
18. Kurkowski, S., Camp, T., Colagrosso, M.: Manet simulation studies: the incredibles. SIGMOBILE Mob. Comput. Commun. Rev. **9**(4), 50–61 (2005). http://doi.acm.org/10.1145/1096166.1096174
19. Laboratory for Communications and Applications (LCA): Traffic and Network Simulation Environment (TraNS). http://lca.epfl.ch/projects/trans/. Accessed 27 May 2013
20. Lacage, M., Henderson, T.R.: Yet another network simulator. In: Proceeding from the 2006 Workshop on ns-2: The IP Network Simulator, WNS2 '06. ACM, New York (2006). http://doi.acm.org/10.1145/1190455.1190467
21. Riley, G.F.: The georgia tech network simulator. In: Proceedings of the ACM SIGCOMM Workshop on Models, Methods and Tools for Reproducible Network Research, MoMeTools '03, pp. 5–12. ACM, New York (2003). http://doi.acm.org/10.1145/944773.944775
22. Riverbed Technology: OPNET. http://www.opnet.com. Accessed 7 June 2013
23. Imre, S., Keszei, C.: Simulation environment for ad-hoc networks in omnet++. In: IST 2001 - "Technologies Serving People" (2001)
24. SCALABLE Network Technologies: QualNet. http://web.scalable-networks.com/content/qualnet. Accessed 7 June 2013
25. The Ohio State University and University of Illinois: The Autonomous Component Architecture. http://j-sim.cs.uiuc.edu/whitepapers/aca.html. Accessed 27 June 2013
26. Transports Metropolitans de Barcelona: Resum de gestió 2011. http://www.tmb.cat/ca/c/document_library/get_file?uuid=c5e4a93e-ed5d-4ad6-bf5a-9d6968f6f375& groupId=10168. Accessed 26 June 2013
27. Zeng, X., Bagrodia, R., Gerla, M.: Glomosim: a library for parallel simulation of large-scale wireless networks. In: Proceedings of the Twelfth Workshop on Parallel and Distributed Simulation, PADS 98, pp. 154–161 (1998)

On the Use of Social Trajectory-Based Clustering Methods for Public Transport Optimization

Jordi Nin[1(✉)], David Carrera[1], and Daniel Villatoro[2]

[1] Barcelona Supercomputing Center (BSC),
Universitat Politècnica de Catalunya (BarcelonaTech), Barcelona, Catalonia, Spain
nin@ac.upc.edu, dcarrera@ac.upc.edu
[2] Barcelona Digital Technology Centre, Barcelona, Catalonia, Spain
dvillatoro@bdigital.org

Abstract. Public transport optimisation is becoming everyday a more difficult and challenging task, because of the increasing number of transportation options as well as the increase of users. Many research contributions about this issue have been recently published under the umbrella of the smart cities research. In this work, we sketch a possible framework to optimize the tourist bus in the city of Barcelona. Our framework will extract information from Twitter and other web services, such as Foursquare to infer not only the most visited places in Barcelona, but also the trajectories and routes that tourist follow. After that, instead of using complex geospatial or trajectory clustering methods, we propose to use simpler clustering techniques as k-means or DBScan but using a real sequence of symbols as a distance measure to incorporate in theclustering process the trajectory information.

Keywords: Smart cities · Geospatial clustering · Metric spaces · OSA distance · Cloud computing · High performance computing

1 Introduction

Trajectory clustering algorithms [24] group similar trajectories into groups (clusters), thus discovering common trajectories. These methods are different from geospatial clustering [38]. This latter clustering methods group similar objects (points in an Euclidian or geodesic space) based on their distance, connectivity, or relative density in the space.

Since geospatial clustering methods are in general easier than trajectory clustering algorithms, they has been employed in the field of spatial analysis for years. Spatial clustering is the process of grouping similar objects based on their distance, connectivity, or relative density in space. Thanks to that, such methods prevail over trajectory clustering algorithms in pattern recognition smart cities applications, such as public transport optimization. The main problem behind

J. Nin and D. Villatoro (Eds.): CitiSens 2013, LNAI 8313, pp. 59–70, 2014.
DOI: 10.1007/978-3-319-04178-0_6, © Springer International Publishing Switzerland 2014

the use of geospatial clustering methods for this purpose is that patterns (most of times trajectories) must be built over geospatial clustering results. To do that, expert knowledge is usually required making this approach less useful for big cities and highly dynamic city environments.

Since trajectory clustering algorithms include by definition all the trajectory information, it is possible to adjust these algorithms to reduce the expert's time to take the decision about which global trajectories are the most interesting ones. In addition, trajectory clustering algorithms can help the data practitioners to discover common sub-trajectories inside a cluster. This information can be very valuable in many applications, especially if we have regions of special interest for analysis inside a big city, such as, the city centre or certain regions with a high number of touristic attractions or city services.

All trajectory clustering algorithms are based on computing a distance assuming that trajectory elements are represented by means of space coordinates. This lack of element semantics makes that clusters construction only considers space similarities. In this paper, we would like to study how to convert trajectory coordinates to symbols, in such a way, we can include semantics inside the trajectory elements. Therefore, it will be possible to group similar trajectories considering the nature of their elements, for instance if they are touristic attractions, city services, restaurants, etc. The main drawback of this approach is that it is required to define an appropriate distance for this type of sequences.

In this paper we propose to use the Optimal Symbol Alignment (OSA) distance [18] for *semantic-aware* trajectory clustering. To do that, we would like to integrate it into the clustering framework ELKI [1] to perform some experiments using trajectories about tourists visiting Barcelona.

1.1 Paper Organization

The rest of this paper is organized as follows. Firstly, in Sect. 2 we introduce some basic concepts about sequences of symbols distances. We also provide a complete definition of the Edit and OSA distances. Then, in Sect. 3 we provide a taxonomy for the currently used clustering algorithms, as well as, some implementation decisions we have already taken. In Sect. 4 we describe how we have obtained the data required for our clustering analysis using several social network services. Later, in Sect. 5, we mention several computational issues we must consider due to the high amount of data available. Finally, Sect. 6 depicts the following steps in our proposal.

2 Distances for Sequences of Symbols

Comparison functions for sequences (of symbols) are important components of many applications, for example clustering, data cleansing and integration.

For this reason, there is a lot of work in computing similarities among sequences of symbols [8, 15, 29]. However, the similarity measures presented in most of those works either do not fulfil the mandatory conditions to be a real distance (most of the times, because the triangular inequality does not hold) or do not contain a proof for that. Formally, a distance function d must satisfy the following properties:

1. Symmetry: $d(A, B) = d(B, A)$ for all sequences A, B
2. Positivity: $d(A, B) \geq 0$ for all sequences A, B
3. Reflexivity: $d(A, A) = 0$ for all sequence A
4. Triangular Inequality: $d(A, B) \leq d(A, C) + d(B, C)$ for all sequences A, B, C

To the best of our knowledge, there are only three sequence measures that fulfil these conditions: the Hamming distance [17], the Levenshtein (Edit) distance [25] and the OSA Distance [18]. The remaining measures are similarity functions instead of real distances because they do not comply with the triangular inequality (or this is not proved). For this reason the application of such measures to the scenarios where having a metric space is a must, such as metric spaces [5], clustering [19] or k-nearest neighbors algorithms [6], becomes unfeasible from a theoretical point of view.

The Hamming and Edit distances present also some problems. For instance, the Hamming distance can only be applied to sequences of the same length, while the Edit distance has a large, both practical and theoretical, complexity $(O(n^2))$. For these reasons many similarity measures have been developed, albeit sacrificing some of the mandatory properties of a distance. For example, the Jaro-Winkler distance [21] is very efficient in terms of practical computational cost when the compared strings are not too large. Therefore, it saves execution time (when compared to Edit distance) in applications where there are many comparisons to be done.

2.1   Edit Distance

The Edit distance [25, 32] measures the difference between two sequences, given by the minimum number of edit operations needed to transform one sequence into the other. An edit operation can be either an insertion, deletion or substitution of a single symbol, although many variations exist in which the set of allowed operations is larger or more restricted. In some way, Edit distance assumes that the differences between two sequences are due to typos or spelling errors.

The Edit distance has found a large variety of applications in many scenarios and has achieved very good results [31]. However, the Edit distance has a large complexity: its computation using classical algorithms [36] based on dynamic programming has a complexity equal to $O(n^2)$, where n is the size of the shortest string. Other algorithms to compute the Edit distance exist [4, 34], having a lower complexity of $O(dn)$, for example, where d is the real Edit distance. Note that, when two completely different strings have to be compared, the complexity of these variants is the same as that of the classical algorithm.

2.2   Optimal Symbol Alignment (OSA) Distance

The intuition behind the Optimal Symbol Alignment (OSA) distance [18] is that strings are close if they have many common symbols, and in addition their common symbols are placed in similar positions, in the strings being compared.

Given a finite alphabet of symbols X, let $A = (a_1, \ldots, a_{n_A})$ and $B = (b_1, \ldots, b_{n_B})$ be two sequences of symbols, where $a_i, b_j \in X$, for $i = 1, \ldots, n_A$, $j = 1, \ldots, n_B$. For any sequence of symbols A, we define as $X_A \subseteq X$ the subset of symbols that appear in A; that is, $X_A = \{x \in X \text{ s.t. } \exists i \in \{1, 2, \ldots, n_A\}$ with $a_i = x\}$. For a symbol $x \in X_A$, we also define the subset of positions $A_x = \{i \in \{1, \ldots, n_A\} \text{ s.t. } a_i = x\}$.

We define the OSA distance $d(A, B)$ between the sequences A and B as

$$d(A, B) = \sum_{x \in X_A \cup X_B} d(x, A, B),$$

where the value $d(x, A, B)$ is defined as

$$d(x, A, B) = \begin{cases} |A_x| & \text{if } x \in X_A - X_B \\ |B_x| & \text{if } x \in X_B - X_A \\ f(x, A, B) & \text{if } x \in X_A \cap X_B \end{cases}$$

Finally, we have to define the value of $f(x, A, B)$, which is the contribution of the symbol x to the distance $d(A, B)$, when this symbol x is included in both sequences A and B. Let us assume without loss of generality that $|A_x| \leq |B_x|$. The idea is to select the subset of $|A_x|$ positions j, from the set B_x, which are globally closest to the set of $|A_x|$ positions in A_x. Namely, if $i_1 < i_2 < \ldots < i_{|A_x|}$ are the positions in A_x, then we select $|A_x|$ positions $j_1 < j_2 < \ldots < j_{|A_x|}$ in B_x minimizing the global distance $|i_1 - j_1| + \ldots + |i_{|A_x|} - j_{|A_x|}|$. We use notation $j_h = \mathtt{pj}(i_h, A, B)$, for $h = 1, \ldots, |A_x|$, to denote the position in B_x that optimally matches position $i_h \in A_x$. We say that j_h is the *projection* of position i_h from sequence A to sequence B. For completeness, we also use the symmetric notation $i_h = \mathtt{pj}(j_h, B, A)$.

Each of these common symbols $a_{i_h} = b_{j_h} = x$, for $h = 1, \ldots, |A_x|$, will contribute with $\frac{|i_h - j_h|}{n_{AB}}$ to the value $f(x, A, B)$, where $n_{AB} = \max\{n_A, n_B\}$. In this way, we ensure that these contributions are bounded by 1. The remaining $|B_x| - |A_x|$ symbols will be considered as non-common symbols, so each of them will contribute with a 1 to the global distance $d(A, B)$.

Taking all these facts into account, we finally have

$$f(x, A, B) = (|B_x| - |A_x|) + \frac{1}{n_{AB}} \sum_{i_h \in A_x} |i_h - \mathtt{pj}(i_h, A, B)| \, .$$

Depending on the differences between the two sequences to be compared (more or less repeated symbols, more or less transpositions, etc.) the OSA distance $d_{\mathrm{OSA}}(A, B)$ will be more or less similar to the Edit distance $d_{\mathrm{Edit}}(A, B)$. But in any case, they will not be very far, because it is easy to prove that $\frac{d_{\mathrm{Edit}}(A, B)}{2} \leq d_{\mathrm{OSA}}(A, B) \leq 2 \cdot d_{\mathrm{Edit}}(A, B)$, for any two sequences A, B.

3   Clustering

Clustering is the task of relating similar elements in a dataset to build groups and create general representations (centroids). It is widely used in many fields as data mining, machine learning, image analysis, etc. Since clustering is a very general concept, there are a great variety of algorithms that apply clustering, focusing, for example, on possible centroids, density, data structures, etc.

Clustering algorithms are divided into different categories: space partitioning, also called top-down methods, hierarchical methods, known as bottom-up methods as well or Density-Based Clustering Methods. Now, we review three well-known clustering algorithms which illustrate these categories: k-means (a space partitioning method), the agglomerative hierarchical clustering and DBScan

k-means Algorithm [26]. It is one of the most commonly used clustering techniques. It is an algorithm to cluster n objects into k partitions $(k < n)$. K-means starts by partitioning randomly the input objects into k initial sets. Then, it calculates the centroid of each set. Following, it constructs a new partition by associating each object with the closest centroid. Finally, the centroids are recalculated for the new clusters. This algorithm is repeated until it convergences, i.e. there is no changes in its centroids. Figure 1 shows an example of k-means over a dataset of 100 gaussian random pairs.

Agglomerative Hierarchical Clustering [20]. This method builds a hierarchy tree, called dendrogram, from the individual elements by progressively merging clusters. Note that, at the beginning each element is considered as an independent cluster. The algorithm starts computing a distance matrix among all the elements to be clustered, where the distance in the (i, j) position corresponds to

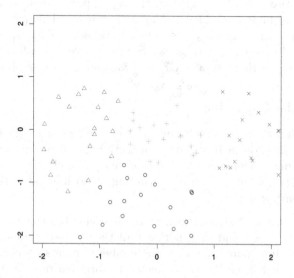

Fig. 1. Example of clustering using k-means algorithm with $k = 5$

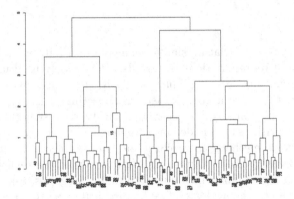

Fig. 2. Example of dendrogram after applying hierarchical clustering

the distance between the ith and jth elements. Then, when clustering progresses, the corresponding rows and columns have to be also merged. This algorithm does not explicitly builds a number of clusters, instead, we must decide the number of clusters and where we split them within the dendrogram. An example, over the same dataset as Fig. 1, is illustrated by Fig. 2, which suggests to cut at height 3 or 4.

DBScan [9]. It starts with an arbitrary starting point r that has not been visited. r's close neighbours are retrieved, and if this set contains sufficiently many points, a cluster is started. Otherwise, r is labeled as noise. Note that this point might later be found in a sufficiently sized environment of a different point and hence be made part of a cluster. If a point is found to be a dense part of a cluster, its neighbours are also part of that cluster. Hence, all points that are found within the neighborhood limit are added to the dense cluster. This process continues until the density-connected cluster is completely found. Then, a new unvisited point is retrieved and processed, leading to the discovery of a further cluster or noise. Figure 3 shows the output of DBScan within a two dense clusters dataset.

3.1 Geospatial Clustering

Geospatial clustering is a special kind of clustering, its main goal is to group similar objects based on their distance, connectivity, or relative density in space. General clustering methods can be used for geospatial clustering, however due to its inherent spatial nature, specific clustering algorithms categories have been defined in the last years:

Grid-Based Clustering Methods. Such methods divide the clustering space into a finite number of cells and then perform all the clustering operations on the grid. Cells containing more than a certain number of points are considered to be dense. Contiguous dense cells are connected to form clusters. One examples of grid-based clustering methods is CLIQUE [2].

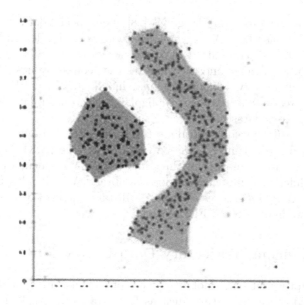

Fig. 3. Example of DBScan with 2 clusters

Constraint-Based Clustering Methods. These clustering methods add spatial constraints, such as obstacles, to obtain more desirable clusters for real geospatial information systems. Depending on the nature of the constraints and applications, constraint-based clustering methods includes four different types of constrains: constraints on individual objects, obstacle objects as constraints, clustering parameters as constraints, and constraints imposed on each individual cluster [33]. DBRS+ [37] is one example of these clustering methods.

3.2   Trajectory Clustering

A number of trajectory clustering methods have been proposed. A common characteristic of first trajectory clustering methods is that they use the shapes of whole trajectories to do the clusters, using for instance hidden Markov models (HMM). Later, more complex clustering algorithms [23] were proposed. In these papers authors divide the space into a grid to allow to compute sub-trajectories and consider more complex patterns than using only the whole trajectory.

Our approach is far from these ideas. Here we would like to adapt classical clustering methods, which has been largely studied to cluster trajectories using well-defined distances for sequences of symbols.

3.3   Clustering Tools

Nowadays, it is possible to find a large variety of clustering tools, such as WEKA [16]. However, most of these tools are not able to work with big data or cannot be parameterized using an external distance.

To solve the first problem, MongoDB [28] offers special data structures and indices called Geohash [27]. When you create a geospatial index on legacy coordinate pairs, MongoDB computes geohash values for the coordinate pairs within the specified location range and then indexes the geohash values. Geohash values, recursively divide a two-dimensional map into quadrants until a desired level of specificity is reach. We have used MongoDB and geohash data structures to create our trajectory dataset.

In order to easily execute several clustering algorithms coming from the different families described in this section, we will use the ELKI framework [1]. ELKI separates data mining algorithms and data management tasks to allow for including different components inside the library. Additionally, ELKI allows for the use of external distances as the OSA distance described in Sect. 2.2. For these reasons we have selected ELKi for our experiments.

4 Social Sensing Trajectory Data Extraction

In order to capture citizens' mobility traces of citizens, different data sources have been used in the literature such as GPS traces [39], call detail records (CDR) from mobile phones [7], and even geopositioned social media [30].

In this work we focus on geopositioned tweets. Twitter allows developers to obtain all the geopositioned tweets within a certain bounding box that covers a certain city. We have used the data-acquisition architecture depicted in [35]. The dataset generated contains the geopositioned tweets generated in Barcelona during six months from July 2012 to the end of December 2012, resulting in approximately 1.160.000 geopositioned data traces.

To transport the geopositioned tweets into actual trajectories, we borrow a methodology presented elsewhere [12]: a trajectory can be understood as the set described by at least two consecutive tweets with a minimum distance of 100 m amongst them and published in less that 75 min of difference.

Moreover, we also profit from a semantically-enhanced module that allows to detect the origin country of each user leaving a digital mobility trace. Each tweet contains the user-specified origin location, allowing users to specify free text values. Other than handling fake values, such as "From heaven", user-specified locations can be referred at different levels of granularity (GPS coordinate, neighborhood, city, region, or country level) and in different languages (e.g. *London, Londra, Londres*, etc...). In order to deal with this ambiguity, we use the GeoNames service [13], which provides an unique country identifier per any input, being also successful dealing with fake values.[1]

Therefore, our technological infrastructure allows us to capture user trajectories and classify them per user origin country.

By applying clustering algorithms on the digital traces of our 6 months dataset, we obtain clusters of activity whose cluster identifier provides us with an unique tag to identify users origins and destinations.

---

[1] A human-supervised test on 100 random user-specified locations obtained from our dataset obtained an 72 % accuracy rate.

5   High Performance Computing Challenges in Geospatial Clustering

To avoid the problems associated to persisting data to disks, persistent memories are leveraged (Flash memory at the current time, any kind of Storage Class Memories in the near future).

To increase the address space made available to a particular workload, the memory of all nodes involved in the execution of such workload are given access to the memory of all other compute nodes through software APIs.

The major problem from the infrastructure point of view for Geospatial Clustering is dealing with large amounts of data. It is a common case that the performance of clustering algorithms is I/O bound because they are data-intensive and process very large data sets. Therefore, novel ways of managing data are required to deliver scalability and performance. The common approach nowadays is to distribute data across many compute nodes and provide applications with a flat and shared name space to access data, independently if it is stored locally or remotely.

The single namespace can be provided either by a distributed file system (the case of MapReduce Distributed File System [3, 14]) or using key/value pairs (the case for most NoSQL databases [22, 28] and key/value stores [11]).

Over the last years an intersection of different technologies is gaining momentum performance-wise: RDMA-enabled network technologies and persistent memories. The goal of combining these technologies is to provide low latency and high bandwidth all-to-all in a network topology, and therefore fast access to all data in a distributed dataset.

High speed networks require lightweight protocol stacks and CPU offloading to move data between nodes at high speeds (e.g. using Infiniband verbs instead of a heavyweight TCP/IP stack to achieve 56 Gbps bandwidth in Infiniband FDR networks), what is achieved using RDMA-enabled networks such as Infiniband or iWARP.

At the same time, to achieve high bandwidth to store data, fast memories are used instead of slow rotational disks. Such memory can be accessed in different ways: through conventional disk interfaces, what is the case of Solid State Disks (SSD); through conventional PCIe buses, what is the case of PCIe Flash boards; or directly through the memory buses in the processors, what is the case of Phase Change Memories (PCM) or any generic Storage Class Memory (SCM) technology.

The outcome of this technology trend is middlewares that allow for transparent access to remote or local data at the same speeds and with the same latencies, unifying the memory space of all nodes involved in the execution of a workload, and simplifying the programmability of the applications by providing simple user APIs, such as the Blue Gene Active Storage [10] (BGAS) platform does.

6   Conclusions

In this abstract we have described all the components of a possible framework for public transport optimisation in big cities, in our case Barcelona. We have described several distances and clustering algorithms to illustrate that combining a distance for sequences of symbols with classical clustering algorithms is possible to create a clustering algorithm for this goal. We have also introduced some performance problems and how we can overcome them. Finally, we have explained how to collect the required data to convert this framework in a real tool during the next year.

Acknowledgments. This work is partially supported by the Ministry of Science and Technology of Spain under contract TIN2012-34557 and by the BSC-CNS Severo Ochoa program (SEV-2011-00067) and with the support of ACC1Ó, the Catalan Agency to promote applied research and innovation; and by the Spanish Centre for Development of Industrial Technology under the INNPRONTA program, project IPT-20111006, "CIUDAD2020".

References

1. Achtert, E., Kriegel, H.-P., Zimek, A.: ELKI: a software system for evaluation of subspace clustering algorithms. In: Ludäscher, B., Mamoulis, N. (eds.) SSDBM 2008. LNCS, vol. 5069, pp. 580–585. Springer, Heidelberg (2008)
2. Agrawal, R., Gehrke, J., Gunopulos, D., Raghavan, P.: Automatic subspace clustering of high dimensional data for data mining applications. SIGMOD Rec. **27**(2), 94–105 (1998)
3. Apache Software Foundation. Hadoop Distributed File System (HDFS) Architecture.
4. Berghel, H., Roach, D.: An extension of Ukkonen's enhanced dynamic programming asm algorithm. ACM Trans. Inf. Syst. **14**(1), 94–106 (1996)
5. Chávez, E., Navarro, G., Baeza-yates, R., Marroquín, J.L.: Searching in metric spaces. ACM Comput. Surv. **33**, 273–321 (1999)
6. Cover, T., Hart, P.: Nearest neighbor pattern classification. IEEE Trans. Inf. Theor. **13**(1), 21–27 (1967)
7. de Montjoye, Y.A., Hidalgo, C.A., Verleysen, M., Blondel, V.D.: Unique in the crowd: the privacy bounds of human mobility. Sci. R. **3**, 1376 (2013)
8. Dong, G., Pei, J.: Sequence Data Mining. Springer, Heidelberg (2007)
9. Ester, M., Kriegel, H., Sander, J., Xu, X.: A density-based algorithm for discovering clusters in large spatial databases with noise. In: 2nd International Conference on Knowledge Discovery and Data Mining, pp. 226–231 (1996)
10. Fitch, B.G., Rayshubskiy, A., Pitman, M.C., Christopher Ward, T.J., Germain, R.S.: Using the active storage fabrics model to address petascale storage challenges. In: Proceedings of the 4th Annual Workshop on Petascale Data Storage, PDSW '09, pp. 47–54. ACM, New York (2009)
11. Fitzpatrick, B.: Distributed caching with memcached. Linux J. **124**, 5 (2004)
12. Gabrielli, L., Rinzivillo, S., Ronzano, F., Villatoro, D.: From tweets to semantic trajectories: mining anomalous urban mobility patterns. In: Proceedings European Conference on Complex Systems ECCS: Barcelona. Springer, Spain (2013)
13. GeoNames. GeoNames geographical database. http://www.geonames.org/ (2010). Accessed July 2013
14. Ghemawat, S., Gobioff, H., Leung, S.-T.: The google file system. SIGOPS Oper. Syst. Rev. **37**(5), 29–43 (2003)

15. Gómez-Alonso, C., Valls, A.: A similarity measure for sequences of categorical data based on the ordering of common elements. In: Torra, V., Narukawa, Y. (eds.) MDAI 2008. LNCS (LNAI), vol. 5285, pp. 134–145. Springer, Heidelberg (2008)

16. Hall, M., Frank, E., Holmes, G., Pfahringer, B., Reutemann, P., Witten, I.H.: The weka data mining software: an update. ACM SIGKDD Explor. Newslett. **11**(1), 10–18 (2009)

17. Hamming, R.W.: Error detecting and error correcting codes. Bell Syst. Tech. J. **26**(2), 147–160 (1950)

18. Herranz, J., Nin, J., Solé, M.: Optimal symbol alignment distance: a new distance for sequences of symbols. IEEE Trans. Knowl. Data Eng. (TKDE) **23**(10), 1541–1554 (2011)

19. Jain, A.K., Murty, M.N., Flynn, P.J.: Data clustering: a review. ACM Comput. Surv. **31**(3), 264–323 (1999)

20. Jardine, N., Sibson, R.: The construction of hierarchic and non-hierarchic classifications. Comput. J. **11**(2), 177–184 (1968)

21. Jaro, M.A.: Advances in record-linkage methodology as applied to matching the 1985 census of Tampa, Florida. J. Am. Stat. Assoc. **84**, 414–420 (1989)

22. Lakshman, A., Malik, P.: Cassandra: a structured storage system on a p2p network. In: Proceedings of the Twenty-First Annual Symposium on Parallelism in Algorithms and Architectures, SPAA '09, p. 47. ACM, New York (2009)

23. Lee, J.G., Han, J., Li, X., Gonzalez, H.: Traclass: trajectory classification using hierarchical regionbased and trajectorybased clustering. In: ACM Very Large Data Base (VLDB) (2008)

24. Lee, J.G., Han, J., Whang, K.Y.: Trajectory clustering algorithms. In: International Conference on Management of Data (SIGMOD), pp. 593–604 (2007)

25. Levenshtein, V.I.: Binary codes capable of correcting deletions, insertions, and reversals. Soviet Phys. Doklady **10**, 707–710 (1966)

26. Lloyd, S.: Least squares quantization in PCM. IEEE Trans. Inf. Theor. **28**, 129–137 (1982)

27. Geohash, http://docs.mongodb.org/manual/core/geospatial-indexes

28. MongoBD, http://www.mongodb.org

29. Navarro, G.: A guided tour to approximate string matching. ACM Comput. Surv. **33**(1), 31–88 (2001)

30. Noulas, A., Scellato, S., Lambiotte, R., Pontil, M., Mascolo, C.: A tale of many cities: universal patterns in human urban mobility. PloS One **7**(5), e37027 (2012)

31. Ristad, E., Yianilos, P.: Learning string edit distance. IEEE Trans. Pattern Recogn. Mach. Intell. **20**(5), 522–532 (1998)

32. Selllers, P.: The theory and computation of evolutionary distances: pattern recognition. J. Algorithms **1**(4), 359–373 (1980)

33. Tung, A.K.H., Han, J., Lakshmanan, L.V.S., Ng, R.T.: Constraint-based clustering in large databases. In: Van den Bussche, J., Vianu, V. (eds.) ICDT 2001. LNCS, vol. 1973, pp. 405–419. Springer, Heidelberg (2000)

34. Ukkonen, E.: On approximate string matching. In: Karpinski, M. (ed.) FCT 1983. LNCS, vol. 158, pp. 487–495. Springer, Heidelberg (1983)

35. Villatoro, D., Serna, J., Rodríguez, V., Torrent-Moreno, M.: The TweetBeat of the city: microblogging used for discovering behavioural patterns during the MWC2012. In: Nin, J., Villatoro, D. (eds.) CitiSens 2012. LNCS, vol. 7685, pp. 43–56. Springer, Heidelberg (2013)

36. Wagner, R.A., Fischer, M.J.: The string-to-string correction problem. J. ACM **21**(1), 168–173 (1974)

37. Wang, X., Hamilton, H.J.: Dbrs: a density- based spatial clustering method with random sampling. In: 7th Pacific-Asia Conference on Knowledge Discovery and Data Mining, pp. 563–575 (2003)

38. Wang, X., Wang, J.: Using clustering methods in geospatial information systems. Geomatica **64**(3), 347–361 (2010)

39. Zheng, Y., Zhang, L., Xie, X., Ma, W.-Y.: Mining interesting locations and travel sequences from gps trajectories. In: Proceedings of the 18th International Conference on World Wide Web, pp. 791–800. ACM, New York (2009)

Migration Movements

Tracking Human Migration
from Online Attention

Carmen Vaca-Ruiz[1,2(✉)], Daniele Quercia[2],
Luca Maria Aiello[2], and Piero Fraternali[1]

[1] Politecnico di Milano, Milan, Italy
{vacaruiz,fraterna}@elet.polimi.it
[2] Yahoo Research, Barcelona, Spain
{dquercia,alucca}@yahoo-inc.com

Abstract. The dynamics behind human migrations are very complex. Economists have intensely studied them because of their importance for the global economy. However, tracking migration is costly, and available data tends to be outdated. Online data can be used to extract proxies for migration flows, and these proxies would not be meant to replicate traditional measurements but are meant to complement them. We analyze a random sample of a microblogging service popular in Brazil (more than 13M posts and 22M reposts) and accurately predict the total number of migrants in 35 Brazilian cities. These results are so accurate that they have promising implications in monitoring emerging economies.

1 Introduction

For census agencies, migrations are difficult to track in the developed countries, let alone in developing ones. In emerging economies, authorities rely on inaccurate, outdated, de-contextualized census data even for the local population [15].

Migrants who have left their home country searching for better opportunities rely also on electronic communication to maintain their bonds with their home communities [3]. Publishing and 'consuming' content such as news and photos in online platforms is "a parcel of everyday life in transnational families" [2]. Previous studies have found that indicators characterizing offline communities (e.g., economic deprivation) can be extracted from online data (e.g., use of emotion words in Twitter) [22]. Therefore, we propose to consider online data in Brazil and track the number of migrants in a city by considering the interaction between users who live in the city and those outside.

Our main contribution is to propose a set of metrics extracted from online data to estimate migration levels. These metrics reflect the intuition that the higher the number of migrants in a city, the more online interactions between users in the city and those outside it. We compute these metrics for 35 cities in a Yahoo Meme dataset that includes more than 13M posts and 22M reposts exchanged between users in more than 1K cities around the world. We find that the proposed metrics work, in that, they correlate with the number migrants reported by the Brazilian census authority. By then combining these metrics in a linear regression model, we show that the model fits the data extremely well (the *Adj.* $R^2 = 0.61$).

J. Nin and D. Villatoro (Eds.): CitiSens 2013, LNAI 8313, pp. 73–83, 2014.
DOI: 10.1007/978-3-319-04178-0_7, © Springer International Publishing Switzerland 2014

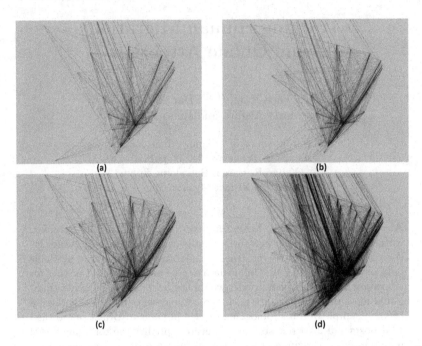

Fig. 1. Evolution of the follower graph: edges connect the geographical locations of the users in our dataset. The picture shows the cumulative set of graph edges after the (a) 1st, (b) 2nd, (c) 3rd and (d) 7th month after the platform launch. The brightest point in (a) corresponds to the city of Sao Paulo.

2   Dataset

Yahoo Meme was a microblogging platform, similar to Twitter, with the exception that users can post content of any length or type (text, pictures, audio, video), being text and pictures the more frequently posted content. In addition to posting, users could also *follow* other users, *repost* others' content, and *comment* on it. In this study, we use a random sample of interactions on Yahoo Meme from its birth in 2009 until the day it was discontinued in 2012 (Table 1). Despite its moderate popularity in USA, Yahoo Meme was popular in Brazil, as witnessed by the fact that the top 45 cities in terms of number of interactions are all located there. Reposting was the main activity in the service (22M sample records) compared to comments (4M). We extract the users who posted the content in our sample and georeference them based on their IP addresses using a Yahoo service. We remove the users for whom we did not obtain results at city level (e.g., users employing proxy servers to connect to the Internet) obtaining 80 K users. For this set of users and their respective posts, we extract all the repost *cascades* and the follower relationships. Month after month, users across different Brazilian cities tended to intensify their follower connections till reaching a certain stability at month 7 after the platform launch (Fig. 1).

Table 1. Yahoo Meme dataset statistics

Property	Value
Number of users	80 K
Number of posts	13M+
Number of reposts	22M+
Number of follower links	19M+
Number of comments	4M
Number of reposts cascades	1.4M
Number of cities	1.3 K
Number repost edges between cities	25 K

To attain geographic representability, we ascertain that the number of users in the top Brazilian cities in our dataset is significantly correlated with the number of Internet users (Fig. 2). As a result, any city outside the confidence area calculated (outlier) is excluded from the study. This leaves us with 35 cities, and we will see that such a number grants statistical significant results. That is because we are left with 1.4M repost cascades whose original content was produced in the 35 cities and was consumed across the world.

3 Attention Metrics

It has been shown that migrants maintain their strong ties in their home countries mainly using digital means [2]. We thus expect that studying online interactions in Yahoo! Meme across geographic areas would result in good estimators of migration flows. More specifically, we connect places every time that a user u_i located in city i interacts with a user located in city j either by reposting u_i's content or by following him/her. The volume of such connections is then correlated with migration rates for 35 cities in Brazil. We consider migrants from Brazil itself and from the rest of the world.

Previous studies have shown that interactions on social media cannot be quantified with simple metrics such as popularity or number of followers but they are best characterized with metrics that also reflect the extent to which content is re-shared or liked [1,6,23,30,32]. That is because social media users make specific decisions about the content they want to consume or who they wish to follow. Such decisions are taken based on offline social ties [31], homophily, and physical distance [25].

We thus resort to attention metrics, and these metrics capture the attention that a city's users are able to attract from the *Rest of the World* and from other *Brazilian* cities:

Cross Border Attention. Our first set of attention metrics for city i is defined as the number of reposts that the city has attracted from the rest of the world (ROW_i^{repost}) or from other Brazilian cities (BR_i^{repost}), normalized with respect to the total number n_i of users in that city:

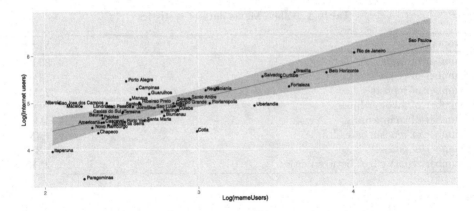

Fig. 2. Number of users in our sample versus number of Internet users. Both quantities are log-transformed. Regression line and 95 % confidence intervals are shown

$$ROW_i^{repost} = \frac{out_i}{n_i}, BR_i^{repost} = \frac{out_i'}{n_i}$$

where out_i is the number of times a post originated in city i has been reposted outside it (the world excluding Brazil); out_i', instead, counts the reposts received outside the city but inside Brazil.

We repeat the same definition considering now the number of cross-borders followers attracted by users in city i:

$$ROW_i^{followers} = \frac{out f_i}{n_i}, BR_i^{followers} = \frac{out f_i'}{n_i}$$

where $out f_i$ is the number of times a user in city i has been followed by a user outside it (the world excluding Brazil); $out f_i'$, instead, counts the follower links outside the city but inside Brazil. As a result, we obtain the first four metrics.

Authority. The previous metrics consider all cities equally. However, certain cities might be more central to migration flows than others. To capture this concept of centrality, we built an *attention graph* using reposts. This is a weighted directed graph where nodes are cities, and directed weighted edges (i, j, w) represent the volume w of reposts between city j where the *reposter* lives, and city i where the original *poster* lives. Self-edges are allowed as many reposts occur between users living in the same city. The resulting *attention graph* has 1,310 nodes and 25 K weighted edges (Fig. 3). Then, we measure the 'authority' index of each city using the HITS algorithm [14]. In the HITS algorithm the autorithy centrality of a vertex is defined to be proportional to the aggregated values of the hub centrality indexes that point to it. For a city i, the two indexes as defined as follows:

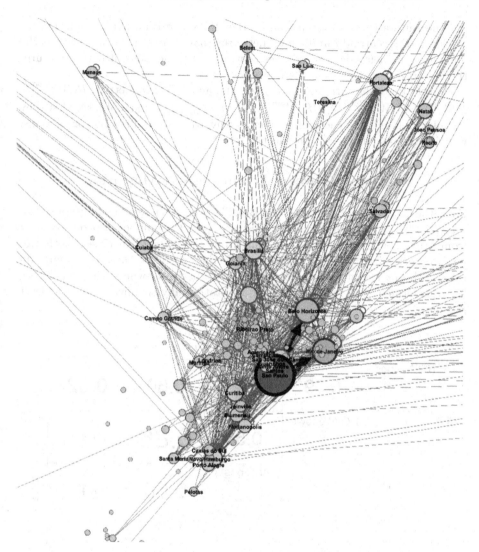

Fig. 3. Attention graph whose nodes are cities and whose weighted edges reflect the intensity of reposting between cities' users. Size and color of the nodes are proportional to the node degree. The network was plotted using the GeoLayout plugin of the Gephi software package [14].

$$Authority_i = \alpha \cdot \sum_{j \in C} A_{ij} Hub_j,$$

$$Hub_i = \beta \cdot \sum_{j \in C} A_{ji} Authority_i,$$

where α and β are constants, C is the set of cities in our dataset and A is the *attention graph*'s corresponding city adjacency matrix.

The Authority index calculated by the HITS algorithm is more informative for the vertex centrality in directed networks than simpler measures such as the number of incident edges or indegree centrality [12] and, thus, it better captures the importance of a node in the network.

We calculate the correlation among each pair of the five metrics: ROW_i^{repost}, BR_i^{repost}, $ROW_i^{followers}$, $BR_i^{followers}$, $Authority_i$ (Fig. 4) and observe that they are all correlated with each other. That is why, when we will run our predictions, we will account for interaction effects.

4   Correlations Between Attention and Migration

From the 2010 data provided by the Brazilian census bureau[1], we compute two *migration rates* for each of the 35 cities: m_{ROW} is the number of people coming from other countries and m_{BR} is that from other Brazilian cities. Both values are normalized by city population. We then correlate these two migration rates with our five attention metrics. To account for skewness, the metrics are log-transformed. The results obtained are statistically significant, with at least p-value < 0.05.

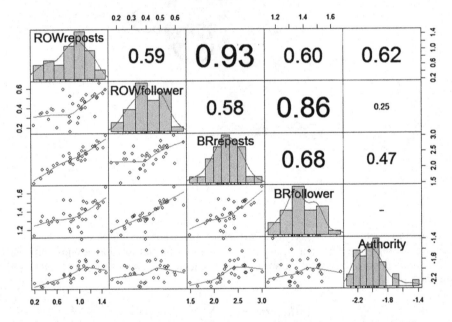

Fig. 4. Correlations among the five attention metrics. We observe that the ROW attention metrics are correlated among each other more than they are with the Authority metric. Values are log-transformed.

---

[1] http://www.ibge.gov.br

Reposts and Follower metrics. We find positive correlations between migration rates and attention received by the rest of the world: $r = 0.28$ for *attention* computed on reposts, and $r = 0.33$ for attention computed on number of followers. Stronger correlations are also found for attention received from other Brazilian cities: $r = 0.33$ for attention computed on reposts, and $r = 0.46$ for attention computed on number of followers.

Authority metric. Since the authority measure can be only computed on the aggregate (Brazil plus rest-of-the-world) dataset, we should correlate the authority measure with the *total* number of migrants ($m_{ROW} + m_{BR}$). In so doing, we obtain, again, a positive correlation $r = 0.32$.

5 Predicting Migration from Attention

We model the number of migrants as a linear combination of the five attention metrics. This is what we call Model1:

$$
\begin{aligned}
log(MigrantsNumber_i) = \alpha + \beta_1 \cdot log(ROW_i^{repost}) + \\
\beta_2 \cdot log(ROW_i^{followers}) + \beta_3 \cdot log(BR_i^{repost}) + \\
\beta_4 \cdot log(BR_i^{followers}) + \beta_5 \cdot log(Authority_i) + \\
\epsilon_i
\end{aligned}
\tag{1}
$$

We also build a model to account for the pairwise interactions effects between indicators:

$$
\begin{aligned}
log(MigrantsNumber_i) = \alpha + \beta_1 \cdot log(ROW_i^{repost}) + \\
\beta_2 \cdot log(ROW_i^{followers}) + \beta_3 \cdot log(BR_i^{repost}) + \\
\beta_4 \cdot log(BR_i^{followers}) + \beta_5 \cdot log(Authority_i) + \\
\gamma_m \cdot Interactions_{im} + \epsilon_i
\end{aligned}
\tag{2}
$$

where $Interactions_{im}$ accounts for the pairwise interactions among the five attention metrics. This is model 2 (Table 2).

To account for Internet penetration rates and population, we build a model adding these two census variables

$$
\begin{aligned}
log(MigrantsNumber_i) = \alpha + \beta_1 \cdot log(ROW_i^{repost}) + \\
\beta_2 \cdot log(ROW_i^{followers}) + \beta_3 \cdot log(BR_i^{repost}) + \\
\beta_4 \cdot log(BR_i^{followers}) + \beta_5 \cdot log(Authority_i) + \\
+\mu_i Internet_i + \rho_i Population_i + \\
+\gamma_m Interactions_{im} + \epsilon_i
\end{aligned}
\tag{3}
$$

where $Internet_i$ is the city's Internet's penetration rate, $Population_i$ is the city's population, and ϵ_i is the error term. This is Model 3. We control for Internet

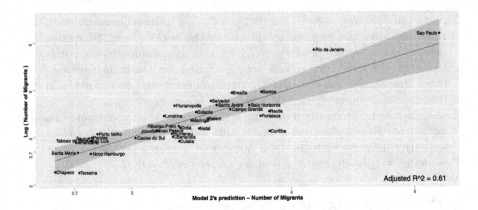

Fig. 5. Predicted values versus actual values calcuated by Model 2 (*Adj.* R^2=0.61) that includes the five attention metrics and their pairwise interactions. The model's prediction error is low: its Mean Absolute Error is 0.21.

Table 2. *Adj.* R^2 for different models predicting city i's number of migrants. Model 1's predictors are the five attention metrics $Attention_{im}$, Model 2 adds their interaction effects, Model 3 controls for the city's Internet penetration rates and population. All *p-values* are < 0.001.

Model	Predictors	$Adjusted.R^2$
1	$\{Attention_{im}\}$	0.54
2	$\{Attention_{im}\} + \{Interactions_{im}\}$	0.61
3	$\{Attention_{im}\} + \{Interactions_{im}\} +$ $Internet_i + Population_i$	0.70

penetration because it is associated with online activity, and for city size because larger cities tend to be economically prosperous and enjoy "increasing returns to scale": a city becomes more attractive as it grows [12].

By computing the beta coefficients of model 2, the one with the best performance (without census data), we find that *cross border attention* in terms of followers accounts for 22 % of the model's explanatory power, while the *cross border attention* for reposts explains 18 %. *Authority* attention, instead, only explains 7 % of the variance. As for model 2's accuracy, the model achieves a Mean Absolute Error (MAE) of 0.21 on a logarithmic scale, where the minimum value is 2.6 and maximum is 5.23, meaning that, on average, the model predicts the log of the number of migrants within 1.16 % of its true value. Figure 5 plots the values predicted by model 2 against actual ones. Rio de Janeiro, one of the most international Brazilian cities, is one outlier for which the number of migrants level is higher than the predicted value.

6   Related Work

Real-life Processes and Social Media. Email exchanges have been used to track migration flows among developed and developing countries [26]. Also, Quercia *et al.* have shown a correlation between the sentiment expressed in tweets originated by residents of London neighborhoods and the neighborhoods' well-being [22].

In the last few years, there have appeared some initiatives for measuring socio-economic conditions of city residents in developing countries using online data. For example, the United Nations and the World Bank have recently launched a program called "Data4Good". This promotes the use of (currently untapped) digital data for, say, improving poverty measurement ("How can we measure poverty more often and more accurately?") or dealing with corruption in international investment projects ("Can we detect fraud by looking at aid data?"). Recently, Orange released an anonymized dataset of mobile phone calls in Côte d'Ivoire, and launched a challenge in which researchers had to predict economic indicators from the activity metrics extracted from the call records [17]. Our research complements this line of work by proposing a set of metrics that can be applied to data extracted from any data source that reflects social exchanges, including social media data.

Migration. Davis *et al.* [8] conducted a study of human mobility using data published by the World Bank. They built a network of countries based on migration flows, and found that the most well connected countries remain stable over time and that migration is directed towards low and mid degree countries.

7   Conclusion

We have shown that online metrics are effective at predicting number of migrants. These metrics are particularly useful in developing countries, where economic changes happen at fast pace. As part of future work, we will study socio-economic indicators other than migration rates, and we will start with GDP and social capital.

Acknowledgments. Carmen Vaca Ruiz's research work has been funded by SENESCYT and ESPOL, Ecuador.

References

1. Asur, S., Huberman, B.A., Szabo, G., Wang, C.: Trends in social media: persistence and decay. In: Proceedings of the 5th AAAI Conference on Weblogs and Social Media (ICWSM) (2011)
2. Baerenholdt, J.O., Granås, B.: Mobility and Place: Enacting Northern European Peripheries. Ashgate Publishing Ltd., Hardcover (2008)
3. Bates, J., Komito, L.: Migration, community and social media. Transnationalism in the Global City, vol. 6. University of Deusto, Bilbao (2012)

4. Boucher, G., Grindsted, A., Vicente, T.L. (eds.): Transnationalism in the Global City. Universidad de Deusto, Bilbao (2012)
5. Brodersen, A., Scellato, S., Wattenhofer, M.: Youtube around the world: geographic popularity of videos. In: Proceedings of the 21st ACM Conference on World Wide Web (WWW) (2012)
6. Cha, M., Haddadi, H., Benevenuto, F., Gummadi, K.: Measuring user influence in twitter: the million follower fallacy. In: Proceedings of the 4th AAAI Conference on Weblogs and Social Media (ICWSM) (2010)
7. Datta, A.: Human Migration: A Social Phenomenon. Mittal Publications, New Delhi (2003)
8. Davis, K.F., D'Odorico, P., Laio, F., Ridolfi, L.: A complex network perspective. PloS One 8(1), e53723 (2013)
9. Eagle, N., Macy, M., Claxton, R.: Network diversity and economic development. Science 328(5981), 1029–1031 (2010)
10. Favell, A., Feldblum, M., Smith, M.P.: The human face of global obility: a research agenda. Society 44(2), 15–25 (2007)
11. Ghosh, R., Lerman, K.: Predicting influential users in online social networks. In: Proceedings of the 4th AAAI Conference on Weblogs and Social Media (ICWSM) (2010)
12. Glaeser, E.L., Kohlhase, J.E.: Cities, regions and the decline of transport costs. Reg., Sci. 83(1), 197–228 (2004)
13. Gruhl, D., Guha, R., Kumar, R., Novak, J., Tomkins, A.: The predictive power of online chatter. In: Proceedings of the Eleventh ACM Conference on Knowledge Discovery in Data Mining (KDD) (2005)
14. Kleinberg, J.M.: Authoritative sources in a hyperlinked environment. J. ACM (JACM) 46(5), 604–632 (1999)
15. Landau, L., Segatti, A.: Contemporary migration to South Africa: a regional development issue. World Bank-free PDF (2011)
16. Lerman, K., Jain, P., Ghosh, R., Kang, J.-H., Kumaraguru, P.: Limited attention and centrality in social networks. In: Proceedings of Conference on Social Intelligence and Technology (SOCIETY) (2013)
17. Mao, H., Shuai, X., Ahn, Y.-Y., Bollen, J.: Mobile communications reveal the regional economy in côte divoire. In: Proceedings of the 3rd Conference on the Analysis of Mobile Phone Datasets (NetMob) (2013)
18. Mejova, Y., Srinivasan, P., Boynton, B.: GOP primary season on twitter: popular political sentiment in social media. In: Proceedings of the Sixth ACM Conference on Web Search and Data Mining (WSDM) (2013)
19. Naaman, M., Becker, H., Gravano, L.: Hip and trendy: characterizing emerging trends on twitter. J. Am. Soc. Inform. Sci. Technol. 62(5), 902–918 (2011)
20. Naveed, N., Gottron, T., Kunegis, J., Alhadi, A.C.: Bad news travels fast: a content-based analysis of interestingness on twitter. In: Proceedings of the Web of Science Conferece (2011)
21. O'Connor, B., Balasubramanyan, R., Routledge, B.R., Smith, N.A.: From tweets to polls: linking text sentiment to public opinion time series. In: Proceedings of the 4th AAAI Conference on Weblogs and Social Media (ICWSM) (2010)
22. Quercia, D., Ellis, J., Capra, L., Crowcroft, J.: Tracking gross community happiness from tweets. In: Proceedings of the ACM Conference on Computer Supported Cooperative Work (CSCW) (2012)

23. Romero, D.M., Galuba, W., Asur, S., Huberman, B.A.: Influence and passivity in social media. In: Gunopulos, D., Hofmann, T., Malerba, D., Vazirgiannis, M. (eds.) ECML PKDD 2011, Part III. LNCS, vol. 6913, pp. 18–33. Springer, Heidelberg (2011)
24. Ruiz, E.J., Hristidis, V., Castillo, C., Gionis, A., Jaimes, A.: Correlating financial time series with micro-blogging activity. In: Proceedings of the Fifth ACM International Conference on Web Search and Data Mining (WSDM) (2012)
25. Scellato, S., Mascolo, C., Musolesi, M., Latora, V.: Distance matters: geo-social metrics for online social networks. In: Proceedings of the 3rd Conference on Online Social Networks (WOSN) (2010)
26. State, W.I., Bogdan, E.Z., et al.: Studying inter-national mobility through ip geolocation. In: Proceedings of the Sixth ACM Conference on Web Search and Data Mining (WSDM) (2013)
27. Taylor, P.J., Ni, P., Derudder, B., Hoyler, M., Huang, J., Lu, F., Pain, K., Witlox, F., Yang, X., Bassens, D., et al.: Measuring the world city network: new developments and results. GaWC Res. Bull. **300** (2009)
28. Tumasjan, A., Sprenger, T.O., Sandner, P.G., Welpe, I.M.: Predicting elections with twitter: what 140 characters reveal about political sentiment. In: Proceedings of the 4th AAAI Conference on Weblogs and Social Media (ICWSM) (2010)
29. UN. Big Data for Development: A Primer. United Nations, Global Pulse (2013)
30. Ver Steeg, G., Galstyan, A.: Information transfer in social media. In: Proceedings of the 21st ACM Conference on World Wide Web (WWW) (2012)
31. Wellman, B., Haase, A., Witte, J., Hampton, K.: Does the Internet Increase, Decrease, or Supplement Social Capital? Social Networks, Participation, and Community Commitment (2001)
32. Weng, J., Lim, E., Jiang, J., He, Q.: Twitterrank: finding topic-sensitive influential twitterers. In: Proceedings of the Third ACM Conference on Web Search and Data Mining (WSDM) (2010)
33. Weng, L., Flammini, A., Vespignani, A., Menczer, F.: Competition among memes in a world with limited attention. Sci. Rep. **2**(335), 1–8 (2012)

Data Anonymization

Beyond Multivariate Microaggregation for Large Record Anonymization

Jordi Nin[✉]

Barcelona Supercomputing Center (BSC),
Universitat Politècnica de Catalunya (BarcelonaTech), Barcelona, Catalonia, Spain
`nin@ac.upc.edu`

Abstract. Microaggregation is one of the most commonly employed microdata protection methods. The basic idea of microaggregation is to anonymize data by aggregating original records into small groups of at least k elements and, therefore, preserving k-anonymity. Usually, in order to avoid information loss, when records are large, i.e., the number of attributes of the data set is large, this data set is split into smaller blocks of attributes and microaggregation is applied to each block, successively and independently. This is called *multivariate microaggregation.* By using this technique, the information loss after collapsing several values to the centroid of their group is reduced. Unfortunately, with multivariate microaggregation, the k-anonymity property is lost when at least two attributes of different blocks are known by the intruder, which might be the usual case.

In this work, we present a new microaggregation method called *one dimension microaggregation* ($Mic1D - k$). With $Mic1D - k$, the problem of k-anonymity loss is mitigated by mixing all the values in the original microdata file into a single non-attributed data set using a set of simple pre-processing steps and then, microaggregating all the mixed values together. Our experiments show that, using real data, our proposal obtains lower disclosure risk than previous approaches whereas the information loss is preserved.

Keywords: Microaggregation · k-anonymity · Privacy in statistical databases

1   Introduction

Managing confidential data is a common practice in any organization. In many cases, these data contain valuable statistical information required by third parties for data analysis and, thus, privacy becomes essential, making it necessary to release data sets preserving the statistics without revealing confidential information. This is a typical problem, for instance, in statistics institutes.

One of the most popular approaches to achieve such a privacy level is to apply *perturbative protection methods* on the microdata source file to be protected. This approach consists in distorting the original data file so that the

J. Nin and D. Villatoro (Eds.): CitiSens 2013, LNAI 8313, pp. 87–107, 2014.
DOI: 10.1007/978-3-319-04178-0_8, © Springer International Publishing Switzerland 2014

resulting data, which is publicly released, does not permit the disclosure of sensitive information. A large number of protection methods exist (see e.g. [1,6,28]). Apart from protecting the privacy of the confidential information, perturbative data protection methods must preserve the statistical utility of the original data as far as possible. In this situation, the main challenge is to find the trade-off between privacy and statistical utility.

Recently, microaggregation has emerged as one of the most promising perturbative data protection methods. For example, Ref. [9] shows that microaggregation is used by many statistical agencies for data anonymization. The basic implementation of microaggregation works as follows [6,7,23]: given a data set with n_{att} attributes, small clusters of at least k elements (records) are built and each original record is replaced with the centroid of the cluster to which the record belongs to. A certain level of privacy is ensured because k records have an identical protected value (k-anonymity [22,25,26]).

However, when n_{att} is large, the statistical utility of the basic microaggregation technique is diminished, specially if the attributes are not highly correlated [2]. This is so because the larger the number of attributes, the larger the distance between the original records in the data set and their corresponding centroids. Therefore, a lot of information on the original data is lost when the protected microdata file is released. To solve this drawback, the following natural strategy is applied by statistical agencies: the microdata file is split into smaller blocks of attributes, and microaggregation is independently applied to each block. This way, the information loss decreases, at the cost of decreasing the achieved level of privacy since the property of k-anonymity is not ensured, as we see later on in this paper. This kind of microaggregation methods are known as *multivariate microaggregation* methods. Another important drawback of this type of methods is that finding the optimal multivariate microaggregation (i.e., finding the clusters that minimize the sum of square errors) is NP-hard [20].

In this work, we propose to combine a set of preprocessing steps along with microaggregation in order to minimize the disclosure risk without losing information. We test this new method using real data showing that $Mic1D-k$ is able to outperform previous multivariate microaggregation methods diminishing the risk of disclosure without increasing the information loss. Specifically, we compare our new method with some of the most commonly used microaggregation methods, showing that $Mic1D-k$ achieves lower disclosure risk than previous algorithms when different groups of attributes are known by an intruder.

This paper is organized as follows. In Sect. 2 we review some basic concepts related to protection methods, focusing on microaggregation techniques. In Sect. 3, we present our new microaggregation method called *One dimension microaggregation*. Section 4 is devoted to compare traditional microaggregation algorithms and our new microaggregation method using real data; we present our experiments and the obtained results. Finally, Sect. 5 draws some conclusions and presents some future work.

2   Preliminaries on Protection Methods

In this section we present some basic concepts that will be useful for the understanding of the work presented in this paper. Namely, we first describe the scenario where a microdata protection method is applied to preserve the privacy of the owners of some statistical data. Then, we explain in detail several microaggregation techniques. Finally, we describe the most usual ways to measure the quality of microaggregation methods, according to the levels of privacy and statistical utility that they provide.

2.1   Statistical Data Protection

Let a microdata file X be a matrix with n rows (records) and n_{att} columns (attributes), where each row contains n_{att} attributes of an individual. The attributes in a microdata file can be classified into two different categories, identifiers or quasi-identifiers, depending on their capability to identify unique individuals. Identifier attributes are used to identify the individual unambiguously. The matrix containing all the values related to these attributes will be denoted in this paper by Id. A typical example of identifier is the passport number. A quasi-identifier attribute is an attribute that is not able to identify a single individual when it is used alone. However, when it is combined with other quasi-identifier attributes, they can uniquely identify an individual. Among the quasi-identifier attributes, we distinguish between confidential (X_c) and non-confidential (X_{nc}), depending on the kind of information they contain. Therefore, we define a microdata file as $X = (Id, X_{nc}, X_c)$. A first naive approach would be to eliminate the identifier attributes and release (X_{nc}, X_c) in order to avoid the linkage of confidential data (X_c) to real individuals. In this scenario, an intruder would be able to re-identify individuals by obtaining the non-confidential quasi-identifier attributes together with identifiers from other data sources and, therefore, disclosing confidential information.

In order to preserve statistical disclosure control, we use assume the solution proposed in [6] to compare several protection methods. This solution is graphically depicted in Fig. 1 and it works as follows:

(i) Identifier attributes in X are either removed or encrypted.
(ii) Confidential quasi-identifier attributes X_c are not modified; in this way, the statistical utility of the confidential attributes is completely preserved.
(iii) A microdata protection method ρ is applied to non-confidential quasi-identifier attributes, in order to preserve the privacy of the individuals whose confidential data is being released, $X'_{nc} = \rho(X_{nc})$.
(iv) The released microdata file is $X' = (\rho(X_{nc}), X_c)$.

In this scenario, as shown in Fig. 2, an intruder might try to re-identify individuals by obtaining the non-confidential quasi-identifier data (X_{nc}) together with identifiers (Id) from other data sources. By applying record linkage between the protected attributes (X'_{nc}) and the same attributes obtained from other data

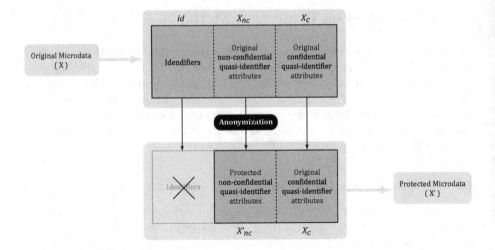

Fig. 1. Data protection and release process.

sources (X_{nc}), the intruder might be able to re-identify a percentage of the protected individuals together with their confidential data (X_c). The quality of a protection methods depends on the percentage of information that it allows to re-identify, among other aspects.

2.2   Microaggregation

As introduced before, microaggregation ensures k-*anonymity* by building small clusters of at least k elements and replacing the original values by the centroid of the cluster to which the record belongs to.

There are other ways to achieve k-anonymity. For instance, in [3] authors present a clustering technique where the released microdata file preserves k-anonymity, as in basic microaggregation. In other solutions, such as those presented in [10], the data holder chooses different subsets of attributes ensuring k-anonymity for each of these subsets independently, similarly to multivariate microaggregation.

We have seen that, in order to solve the information loss problem of the basic microaggregation method, *multivariate microaggregation* is used at the cost of increasing the disclosure risk. Specifically, after dividing attributes into different blocks and applying the basic microaggregation technique to each block separately, the k records which fall in the same cluster for the first block of attributes, may fall in a different cluster for any of the other blocks of attributes. So, the resulting protected records will not be equal and no k-anonymity is ensured. The easiest case for microaggregation in terms of attribute blocking complexity occurs when the size of the attribute blocks is equal to one. In other words, when each attribute is protected independently. This corresponds to *Univariate Microaggregation* or *Individual Ranking Microaggregation*.

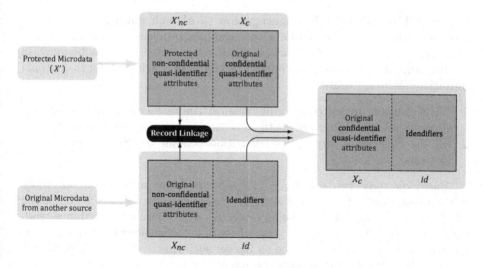

Fig. 2. Disclosure risk scenario.

In order to preserve information loss as low as possible, microaggregation methods try to minimize the total sum of distances between all the elements to be protected and the centroid of the cluster where an element belongs to, i.e minimize the total Sum of Square Errors (SSE):

$$SSE = \sum_{i=1}^{c} \sum_{x_{ij} \in C_i} (x_{ij} - \bar{x}_i)^T (x_{ij} - \bar{x}_i),$$

where c is the total number of clusters, C_i is the ith cluster and $\bar{x}_i$ is the centroid of C_i. The restriction is $|C_i| \geq k$, for all $i = 1, \ldots, c$. In general, the larger value of k the lower the disclosure risk. Therefore, in order to parametrize microaggregation methods, k has to be as large as possible without compromising the statistical utility of the protected information.

The rational of this process is to make the protected data as similar as possible to the original one. In any case, methods should provide clusters with at least k elements. As introduced before, finding the optimal multivariate microaggregation has been proven to be an NP-Hard problem. For this reason, heuristic methods have been proposed in the literature.

For our work, we will use some of these different algorithms proposed for microaggregation in order to compare them to our new microaggregation technique. In this section, we explain a deterministic and optimal algorithm for univariate microaggregation, which is also used by two other methods for projection based multivariate microaggregation: PCP microaggregation and Zscores microaggregation. Finally, we describe one of the most used methods for heuristic microaggregation (specially for the multivariate case, although it can be applied to the univariate case as well): the MDAV (Maximum Distance to Average Vector) algorithm.

Optimal Univariate Microaggregation. Although multivariate microaggregation has been proven to be a very complex problem, several polynomial approaches for the optimal univariate microaggregation as [11] may be found in the literature. In [7], the authors present two relevant conclusions for the optimal univariate microaggregation:

1. When elements are sorted according to an attribute, for any optimal partition, elements in each cluster are contiguous (non overlapping clusters exist)
2. All the clusters of any optimal partition contain between k and $2k - 1$ elements.

Based on these two results, in [11] authors define an optimal univariate microaggregation as follows. Let $A = (v_1 \ldots v_n)$ be a vector of size n containing all the values for the attribute being protected. The values are sorted in ascending order so that if $i < j$ then $v_i \leq v_j$, where v_1 is the smallest element and v_n is the largest element in A. Let k be an integer such that $1 \leq k < n$ (k is directly obtained from the microaggregation configuration).

Given A and k, a graph $G_{k,n}$ is defined as follows. Firstly, we define the nodes of G as the elements v_i in A plus one additional node g_0 (this node is later needed to apply the Dijkstra algorithm). Then, for each node g_i, we add to the graph the directed edges (g_i, g_j) for all j such that $i + k \leq j < i + 2k$. The edge (g_i, g_j) means that the values $(v_i, \ldots, v_j)$ might define one of the possible clusters. Then, the cost of the edge (g_i, g_j) is defined as the within-group sum of squared error for such cluster. That is, $SSE = \Sigma_{l=i}^{j}(v_l - \bar{v})^2$, where $\bar{v}$ is the average record of the cluster.

Given this graph, the optimal univariate microaggregation is defined by the shortest path algorithm between the nodes g_0 and g_n. This shortest path can be computed using the Dijkstra algorithm. Thus, the optimal clustering can be computed in linear time.

Projection Based Microaggregation. The basic idea of Projection Based Microaggregation methods is to approximately reduce the multivariate microaggregation problem into the univariate case, by projecting $n_{att} > 1$ attributes (corresponding to some attributes of the records) into a single one.

The use of this projection techniques is motivated by the difficulty of sorting multivariate data that arises when one tries to extend the optimal univariate solution to the case of multivariate microaggregation.

Ideally, the employed projection should maintain the global statistical properties of the initial (non-projected) values. With this goal in mind, two projection methods seem particularly suitable: the principal component projection [12] and the sum of Z-scores [13].

Projected multivariate microaggregation is described in Algorithm 1, when applied to a microdata file X with n records and n_{attr} attributes.

Ideally, the employed projection should maintain the global statistical properties of the initial (non-projected) values. With this goal in mind, two projection methods seem particularly suitable: the principal component projection (PCP)

Algorithm 1: Projected Microaggregation

Data: X: original microdata, k: integer
Result: X': protected microdata
begin

Split the microdata file X into r sub-files $\{X_i\}_{1 \leq i \leq r}$, each one with v_i
attributes of the n records, such that $\sum_{i=1}^{r} v_i = V$;

foreach $(X_i \in X)$ **do**

Apply a projection algorithm to the attributes in X_i, which results in
an univariate vector z_i with n components (one for each record) ;
Sort the components of z_i in increasing order;
Apply to the sorted vector z_i the following variant of the univariate
optimal microaggregation method explained in Sect. 2.2: use the
algorithm defining the cost of the edges $\langle z_{i,s}, z_{i,t} \rangle$, with $s < t$, as the
within-group sum of square error for the v_i-dimensional cluster in X_i
which contains the original attributes of the records whose projected
values are in the set $\{z_{i,s}, z_{i,s+1}, \ldots, z_{i,t}\}$;
For each cluster resulting from the previous step, compute the
v_i-dimensional centroid and replace all the records in the cluster by the
centroid ;

end

---

[12] and the sum of Z-scores [13]. PCP is a projection technique that preserves the variance of the multivariate data set as much as possible in the projected data set in order to simplify the complexity of the data set, while preserving the statistical utility of the projected data. On the other hand, Z-score is a dimensionless quantity derived by subtracting the mean of each attribute from a single value and then dividing the difference by the standard deviation of that attribute. The resulting microaggregation algorithms obtained after the application of these two projection methods, called *PCP microaggregation* and *Zscores microaggregation*, will be used in Sect. 4 to test the quality of our new technique. See [17] for more details.

MDAV Microaggregation. The MDAV (Maximum Distance to Average Vector) algorithm [7,14] is an heuristic algorithm for clustering records in a microdata file X so that each cluster is constrained to contain at least k records. This algorithm can be used for univariate microaggregation and multivariate microaggregation. The MDAV algorithm is described in Algorithm 2.

MDAV generic algorithm can be instantiated for different data types, using appropriate definitions for distance and average. Normally, *the most distant record* and the *closest records* are computed using the Euclidean distance, and *the average record* is defined as the arithmetic mean of the records. The average record is used to replace the original records when building the protected microdata file.

Algorithm 2: MDAV

Data: X: original microdata, k: integer
Result: X': protected microdata
begin
    while ($|X| > k$) **do**
        Compute the average record $\bar{x}$ of all records in X;
        Consider the most distant record x_r to the average record $\bar{x}$;
        Form a cluster around x_r. The cluster contains x_r together with the
        $k-1$ closest records to x_r;
        Remove these records from microdata file X;
        if ($|X| > k$) **then**
            Find the most distant record x_s from record x_r;
            Form a cluster around x_s. The cluster contains x_s together with the
            $k-1$ closest records to x_s;
            Remove these records from microdata file X;
        Form a cluster with the remaining records;
end

2.3 Measures to Evaluate Risk and Utility

As we discussed before, a microdata protection method must guarantee a certain level of privacy (low disclosure risk). At the same time, since the goal is to allow third parties to perform reliable statistical computations over the released (protected) data, the protection method must ensure that the protected data is statistically close enough to the original one.

Therefore, given a microdata protection method, we have two inversely related aspects to measure: the *disclosure risk* (*DR*), which is the risk that an intruder obtains correct links between the protected and the original data; and the *information loss* (*IL*) caused by the protection method. When one of them increases, the other one decreases. The two extreme cases are the following ones: (i) if the original microdata is released, then information loss is zero, but the disclosure risk is maximum; (ii) if the original microdata is encrypted and then released, the disclosure risk is (almost) zero, but the information loss is maximum.

There are different generic measures proposed in the literature to evaluate the quality of a data protection method. One approach was presented in [5,6], where the authors combine both information loss and disclosure risk in a *score* using the arithmetic mean. This method is refined in subsequent works [19,24].

In order to calculate the *score*, first we need to calculate some information loss and disclosure risk meausures:

– **Information Loss (IL):** Let X and X' be matrices representing the original and the protected microdata files, respectively. Let V and R be the covariance matrix and the correlation matrix of X, respectively; let $\overline{X}$ be the vector of variable averages for X and let S be the diagonal of V. Define V', R', $\overline{X}'$, and S' analogously from X'. The information loss is computed by averaging the mean variations of $X - X', V - V', S - S'$, and the mean absolute error of $R - R'$ and multiplying the resulting average by 100.

- **Disclosure Risk (DR):** For this measure in the original paper, the authors assume two different scenarios in order to evaluate DR: (i) *Distance Linkage Disclosure risk* (DLD), which is the average percentage of linked records using distance based RL [21]. Note that, this scenario is the same described in Sect. 2.1, where the intruder has access to an external data source and he is interested in disclosure the identity of the individual and (ii) *Interval Disclosure risk* (ID) which is the average percentage of original values falling into the intervals around their corresponding masked values, in this scenario the intruder has no access to an external data source and he is interested to disclose the original value of a protected one. The two values are computed over the number of attributes that the intruder is assumed to know, in particular, in this paper we assume the scenario described in [19] where the intruder knows all the possible combinations from one to all the attributes. The Disclosure Risk is computed as $DR = 0.5 \cdot DLD + 0.5 \cdot ID$.

- **Score:** The final score measure is computed by weighting the presented measures and it was also proposed in [6]:

$$score = 0.5\, IL + 0.5\, DR$$

Note that the better a protection method, the lower its score.

Apart from this generic measure for protection method evaluation, we can find specific IL and DR measures for microaggregation in the literature. For instance, the total Sum of Square Error SSE is usually used for information loss evaluation, since it is the fitness function used by microaggregation to minimize the loss of statistical utility of the protected microdata file. In order to compute the DR, in [18], a specific DR measure is defined. The idea is to consider the ratio between the total number n of records and the number of protected records which are different. This gives the average size of each 'global cluster' in the protected microdata file. This measure was denoted as k', this *real anonymity* measure is computed as

$$k' = \frac{n}{|\{x'|x' \in X'\}|}$$

Since our work compares our new microaggregation algorithm with other classical microaggregation methods, we use both the specific measures presented above and the general *score* measure.

3   One Dimension Microaggregation

In this section, we present a new microaggregation method called *one dimension microaggregation* (Mic1D-k, for short). This method gathers all the values of the microdata file into a single sorted vector, independently of the attribute they belong to. Then, it microaggregates all the mixed values together. The experiments presented here show that, by using real data, our proposal obtains

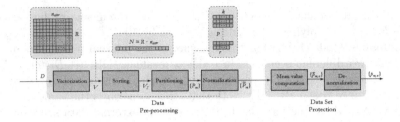

Fig. 3. Mic1D-k schema.

lower disclosure risk than previous approaches whereas the information loss is preserved.

As shown in Fig. 3, Mic1D-k is based on an important data pre-processing technique that must be applied before starting the protection process. This pre-processing phase is decomposed in several steps. Namely, vectorization, sorting, partitioning and normalization. Following, we go into further details about these steps.

Vectorization

The vectorization step gathers all the values from the microdata file in a single vector, independently on the attribute they belong to. Thereby, we ignore the attribute semantics and therefore the possible correlation between two different attributes in the microdata file. In other words, we *desemantize* the microdata file. Later, this process plays a central role in the discussion about the results achieved by Mic1D-k.

Formally speaking, let $\mathcal{D}$ be the original microdata file to be protected. We denote by R the number of records in $\mathcal{D}$. Each record consists of n_{att} numerical attributes. We assume that none of the records contains missing values. We denote by N the total number of values in $\mathcal{D}$. As a consequence, $N = R \cdot n_{att}$.

Let V be a vector of size N containing all the values in the microdata file. Mic1D-k treats values in the microdata file as if they were completely independent. In other words, the concept of record and attribute is ignored and the N values in the microdata file are placed in V.

The effect of this step on a certain microdata file is depicted in the upper half of Fig. 3.

Sorting

Since the values in the vectorized microdata file belong to different source attributes, they present a pseudo-random aspect and it becomes very difficult to find the optimal partitions, i.e. partitions with SSE value as low as possible. In order to simplify this search, the whole vector is sorted. This way, by the conclusions extracted from the univariate microaggregation presented in Subsect. 2.2, optimal partitions are contiguous and, therefore, the partitioning process in this new vector can be done easily, as we will see later.

Formally, V is sorted increasingly. Let us call V_s the sorted vector of size N containing the sorted data and v_i the ith element of V_s, where $0 \leq i < N$.

Partitioning

Similarly to general microaggregation, in order to ensure a certain level of privacy (k-anonymity), Mic1D-k splits the vectorized microdata file in several k-partitions and it calculates the average value for each partition. By modifying the value of k, Mic1D-k allows us to adjust the trade-off between information loss (SSE) and disclosure risk. Note that if the vectorized microdata file was not sorted (previous step), k would not have this property.

Formally, V_s is divided into smaller sub-vectors or partitions. We define k where $1 < k \leq N$ as the number of values per partition. Note that if k is not a divisor of N, the last partition will contain a smaller number of values. Let P be the number of partitions containing k values. We call r the number of values in the last partition where $0 \leq r < k$. Therefore, $N = kP + r$. We will suppose that $r > 0$, so we have $P + 1$ partitions (note that $r > 0$ if and only if k does not divide N). We denote by P_m the mth partition.

Let $v_{m,n}$ be defined as the nth element of P_m:

$$\begin{cases} v_{m,n} := v_{mk+n} & n = 0 \ldots k - 1 \quad m = 0 \ldots P - 1 \\ v_{P,n} := v_{Pk+n} & n = 0 \ldots r - 1 \end{cases}$$

The upper half of Fig. 3 shows the effect of this step on a certain microdata file.

Normalization

Since the range of the values in the different attributes could differ significantly among them, it is necessary to normalize the data to a certain predefined range of values.

There are many ways to normalize a microdata file. A possible solution would be to normalize each attribute independently before the application of the vectorization step. However, this normalization method could present problems with skewed attributes and therefore the attributes could not be merged in the sorting step. For this reason, we propose to normalize the data stored in each partition separately. Thereby, similar values are assigned to the same partition and therefore the chances to avoid the effect of skewness in the data are higher.

Formally, we denote the normalized values as $\bar{v}_{m,n}$ and the normalized partitions as $\bar{P}_m$. Let $\max_m$ and $\min_m$ be the maximum and the minimum values in the mth partition:

$$\max_m := \max_{0 \leq i < k} \{v_{m,i}\} \qquad \min_m := \min_{0 \leq i < k} \{v_{m,i}\}$$

The normalized values are then defined as:

$$\begin{cases} \bar{v}_{m,n} := \frac{v_{m,n} - \min_m}{\max_m - \min_m} & \text{if } \max_m \neq \min_m \\ \bar{v}_{m,n} := 0.5 & \text{if } \max_m = \min_m \end{cases}$$

where $0 \leq m < P$ (or $0 \leq m \leq P$ if k does not divide N,) and $0 \leq n < k$. Note that $\max_m = \min_m$ means that all the values in the partition are the same. In this case, the normalized value is centered in the normalization range.

Re-sorting and Re-normalization

One of the goals of the sorting process, apart from reducing the SSE value, is to desemantize the microdata file, i.e., to merge values from different attributes in order to completely break their semantics and therefore make the re-identification process more difficult. If the range of values of a certain attribute differs significantly from the others, it is likely that it is not merged in previous steps.

Then, in order to appropriately mingle all attributes, once data has been sorted and normalized, we repeat these two steps (sorting and normalization). Since the range of values have been homogenized by normalization, attributes are conveniently mixed in the second sorting step and thus the microdata file is correctly preprocessed.

Mean Value Computation

Once data is preprocessed, for each partition $\bar{P}_m$, the mean value of its components is computed:

$$\mu_m = \sum_{n=0}^{k-1} \frac{\bar{v}_{m,n}}{k} \qquad m = 0 \ldots P - 1 \qquad \mu_P = \sum_{n=0}^{r-1} \frac{\bar{v}_{P,n}}{r}$$

where the latter expression is applied to the last partition if $r > 0$, i.e., if k does not divide the total number of values in the microdata file.

The protected value $\bar{p}_{m,n}$ for $\bar{v}_{m,n}$ is then:

$$\begin{cases} \bar{p}_{m,n} = \mu_m & n = 0 \ldots k - 1 \quad m = 0 \ldots P - 1 \\ \bar{p}_{P,n} = \mu_P & n = 0 \ldots r - 1 \end{cases}$$

Finally, Mic1D-k denormalizes the data into the original range, according to the normalization and re-normalization steps in the previous block. Then, the protected values are placed in the protected microdata file in the same place occupied by the corresponding $v_{m,n}$ in the original microdata file. In this way, we are undoing the sorting and vectorization steps.

4   Experiments

We have tested Mic1D-k with real data extracted from two microdata files available in the Internet. The first one, denoted as Water-treatment, was extracted from the UCI repository [15], and it has already been used in other works dealing with disclosure risk evaluation, e.g. [16]. It contains 35 attributes corresponding to 380 entries or records. The second microdata file, called Census, was extracted using the Data Extraction System [27] of the U.S. Census Bureau, and it has been used as a reference database in many works dealing with statistical data protection, e.g. [4,6,17]. It contains 1080 records and 13 attributes.

As we will see later on, Mic1D-k achieves lower disclosure risk using real data than other multivariate microaggregation methods, such as MDAV, whereas it preserves a lower information loss.

4.1   Attribute Selection

To apply multivariate microaggregation to a microdata file X, we need to choose among the different microaggregation methods, the parameter k, and the number of blocks the microdata file X is split into. However, there are other parameters to be considered when the number of blocks B is larger than 1. As it was explained in [18], the way in which the attributes are grouped into blocks affects significantly the results and the quality of multivariate microaggregation.

It is a standard practice in statistical agencies to select the attributes on the basis of statistical utility. It is clear that, if the considered attributes are highly correlated, two records which are similar with respect to one attribute, will be also similar with respect to another one. Due to this, if microaggregation is applied to correlated attributes, when two values of the same attribute coming from different records are close, each pair of attributes coming from the remaining attributes in the cluster will be also close. Then, the intra-cluster distance is short and the information loss is low.

Nevertheless, as usual, statistical utility and privacy are inversely related. Therefore, the disclosure risk of microaggregation in this case is higher than in the case where correlated attributes are put into different blocks. Then as pointed out in [18], it is possible to group the attributes in a different way: blocks are formed in such a way that the first attributes of all blocks are (highly) correlated, the second attributes of all blocks are also (highly) correlated, and so on. This way, we are making blocks correlated, instead of constructing blocks with correlated attributes. The goal of this approach is to increase the resulting real anonymity k'. If two records A and B are in the same cluster for some blocks, this means that the first attribute values of these records are more or less close to each other, and the same for the second attribute of the block, etc. Then, when we consider another block, if the jth attribute of this new block is (highly) correlated with the jth attribute of the latter block, records A and B will probably be close to each other as well, with respect to the attributes in the second block. Therefore, with some non-negligible probability, A and B will fall in the same cluster, again. Ideally, some records will fall inside the same clusters, for each block of attributes, and so the number of protected records which will be exactly equal will be higher, increasing in this way the real anonymity and the privacy level of the released data.

4.2   Algorithms Parameterization

We have tested Mic1D-k and compared our results with those obtained by the projected microaggregation (PCP and Zscores) and MDAV microaggregation, using the Census and Water Treatment microdata files. As we explained above, when protecting a microdata file using multivariate microaggregation, the way in which the data is split to form blocks is highly relevant with regard to the degree of privacy achieved (k' value). For this reason, we have reduced both microdata files to have 9 attributes, which we detail in Tables 1 and 2.

In both files, attributes $a1, a2$ and $a3$ are highly correlated as well as attributes $a4, a5$ and $a6$ and attributes $a7, a8$ and $a9$. On the contrary, attributes in different blocks (e.g. $a1$ and $a4$) are non-correlated. For our experiments, when protecting data, we assume attributes to be split into three blocks of three attributes each. Also, we consider two situations when protecting the microdata files: blocking correlated attributes and, therefore, having non-correlated blocks (low information loss and low disclosure risk), i.e., $(a1, a2, a3)$, $(a4, a5, a6)$ and

Table 1. Attribute description of the water-treatment microdata file.

id	Name	Description
$a1$	PH-E	Input pH to plant
$a2$	PH-P	Input pH to primary settler
$a3$	PH-D	Input pH to secondary settler
$a4$	DQO-E	Input chemical demand of oxygen to plant
$a5$	COND-P	Input conductivity to primary settler
$a6$	COND-D	Input conductivity to secondary settler
$a7$	DBO-S	Output biological demand of oxygen
$a8$	SS-S	Output suspended solids
$a9$	SED-S	Output sediments

Table 2. Attribute description of the census microdata file.

id	Name	Description
$a1$	AGI	Adjusted gross income
$a2$	FICA	Social security retirement payroll deduction
$a3$	INTVAL	Amount of interest income
$a4$	EMCONTRB	Employer contribution for health insurance
$a5$	TAXINC	Taxable income amount
$a6$	WSALVAL	Amount: Total wage and salary
$a7$	ERNVAL	Business or farm net earnings in 19
$a8$	PEARNVAL	Total person earnings
$a9$	POTHVAL	Total other persons income

$(a7, a8, a9)$; and blocking non-correlated attributes but correlated blocks, i.e., $(a1, a4, a7)$, $(a2, a5, a8)$ and $(a3, a6, a9)$.

For each microdata file and attribute selection method, we apply all microaggregation methods using different configurations (i.e. different values of k). The selection of these values aims at covering a wide range of SSE values and, thus, studying scenarios with different *information loss* values. Namely, we protect the microdata files with parameter $k = 5, 25, 50$ for the Census microdata file, and $k = 5, 15, 25$ for the Water-treatment microdata file.

Table 3. Different groups of attributes known by the intruder.

Correlated	1G	$(a1, a2, a3)$, $(a4, a5, a6)$, $(a7, a8, a9)$
	2G	$(a1, a2, a5)$, $(a1, a3, a7)$, $(a2, a3, a6)$, $(a1, a4, a5)$, $(a2, a4, a6)$ $(a5, a6, a9)$, $(a6, a7, a8)$, $(a1, a8, a9)$, $(a2, a7, a9)$
	3G	$(a1, a4, a7)$, $(a1, a5, a8)$, $(a1, a6, a9)$, $(a2, a4, a7)$, $(a2, a5, a8)$ $(a2, a6, a9)$, $(a3, a4, a7)$, $(a3, a5, a8)$, $(a3, a6, a9)$
Non-correlated	1G	$(a1, a4, a7)$, $(a2, a5, a8)$, $(a3, a6, a9)$
	2G	$(a1, a4, a5)$, $(a1, a3, a7)$, $(a4, a7, a8)$, $(a1, a2, a5)$, $(a2, a4, a8)$ $(a5, a8, a9)$, $(a3, a6, a8)$, $(a1, a6, a9)$, $(a3, a4, a9)$
	3G	$(a1, a2, a3)$, $(a1, a5, a6)$, $(a1, a8, a9)$, $(a2, a3, a4)$, $(a4, a5, a6)$ $(a4, a8, a9)$, $(a2, a3, a7)$, $(a5, a6, a7)$, $(a7, a8, a9)$

For Mic1D-k, we use $k = 3000, 4000, 5000$ for the Census microdata file and $k = 500, 800, 900$ for the Water Treatment microdata file. Note that, since Mic1D-k *desemantizes* the microdata file, it does not make sense to consider different situations related to the correlation of the attributes and, therefore, we protect the data just once for each parametrization. In order to make a fair comparison, we have chosen the values of k in Mic1D-k to obtain similar SSE values to those obtained by MDAV after protecting the microdata files.

4.3   Algorithms Comparison

In order to compare the disclosure risk of microaggregation methods, we consider that a possible intruder knows the values of three random attributes of the original microdata file. Different tests are performed assuming that the intruder knows different sets of three attributes. Depending on these attributes, by using multivariate microaggregation, the intruder will have information coming from one or more groups. Table 3 shows all the considered possibilities.

Firstly, we suppose that the three known attributes belong to the same microaggregated block (e.g. $(a1, a2, a3)$ in the correlated scenario or $(a1, a4, a7)$ in the non-correlated). Since the size of the three microaggregation blocks is 3, there are only three options to consider. We denote this case by 1G. Since

Table 4. SSE and real k' of different microaggregation methods and parameterizations using the Census microdata file. Method-k corresponds to microaggregation using method Method (MDAV, PCP or Zscores) with initial anonymity value k.

Method	k	SSE	k' 1G	k' 2G	k' 3G
MDAV-k	5	64.99	5.00	1.92	1.00
	25	223.73	25.12	7.00	1.09
	50	328.31	51.43	14.66	1.41
PCP-k	5	131.05	5.06	1.91	1.00
	25	320.76	25.12	6.72	1.02
	50	441.66	51.43	13.97	1.15
Zscores-k	5	66.62	5.05	1.91	1.00
	25	159.95	25.12	6.80	1.04
	50	243.81	51.43	14.23	1.30
Mic1D-k	3000	32.27	8.37	9.87	5.77
	4000	129.06	20.10	22.09	13.89
	5000	738.12	72.83	76.08	55.02

Correlated attributes

Method	k	SSE	k' 1G	k' 2G	k' 3G
MDAV-k	5	58.49	5.00	1.96	1.02
	25	260.13	25.12	7.35	1.24
	50	356.47	51.43	15.86	2.05
PCP-k	5	124.99	5.03	1.91	1.00
	25	251.53	25.12	6.74	1.03
	50	382.69	51.43	14.00	1.22
Zscores-k	5	121.68	5.04	1.93	1.00
	25	242.26	25.12	6.97	1.07
	50	354.83	51.43	14.86	1.45
Mic1D-k	3000	32.27	5.63	8.51	8.04
	4000	129.06	13.53	19.45	19.19
	5000	738.12	59.77	67.77	67.25

Non-correlated attributes

Table 5. SSE and real k' of different microaggregation methods and parameterizations using the Water Treatment microdata file. Method-k corresponds to microaggregation using method Method (MDAV, PCP or Zscores) with initial anonymity value k.

Method	k	SSE	k' 1G	k' 2G	k' 3G
MDAV-k	5	28.18	5.09	1.94	1.00
	15	72.03	15.20	4.42	1.01
	25	114.56	25.33	7.28	1.10
PCP-k	5	28.59	5.14	1.94	1.01
	15	71.99	15.20	4.41	1.02
	25	110.91	25.33	7.24	1.04
Zscores-k	5	23.78	5.14	1.94	1.01
	15	72.23	15.20	4.43	1.03
	25	111.69	25.33	7.23	1.07
Mic1D-k	500	65.89	3.25	3.39	1.76
	800	80.95	7.87	7.55	4.67
	900	255.64	12.95	13.61	9.14

Correlated attributes

Method	k	SSE	k' 1G	k' 2G	k' 3G
MDAV-k	5	69.51	5.00	2.03	1.03
	15	173.96	15.20	5.28	1.39
	25	259.07	25.33	9.22	1.91
PCP-k	5	93.67	5.02	1.94	1.01
	15	170.12	15.20	4.47	1.03
	25	229.50	25.33	7.33	1.13
Zscores-k	5	73.52	5.02	1.97	1.02
	15	160.30	15.20	4.75	1.10
	25	231.81	25.33	8.21	1.46
Mic1D-k	500	65.89	2.78	2.58	2.63
	800	80.95	4.74	7.17	6.88
	900	255.64	9.07	14.52	11.71

Non-correlated attributes

the intruder has only access to data from one group, multivariate microaggregation ensures the k-anonymity property (this is the best possible scenario for multivariate microaggregation). However, note that, usually, the intruder cannot choose the attributes obtained from external sources and it might be difficult to obtain all the attributes for the same group. Secondly, we assume that the known attributes belong to two different microaggregated groups. There are many possible combinations of three attributes under this assumption, so nine of them were chosen randomly. We refer to this case as 2G. Finally, case 3G is defined analogously to 2G, and also nine possibilities of known attributes are considered. Note that, in both scenarios 2G and 3G, k-anonymity is not ensured by multivariate microaggregation. Note also that, if the intruder had more than three attributes, it would not be possible to consider 1G. We are considering the case were the intruder only has three attributes to study a scenario were multivariate microaggregation can still preserve k-anonymity.

The second column of Tables 4 and 5 presents the SSE values for all the parameterizations and situations described before. Note that the range of SSE covered by all the methods is similar. This allows us to compare the disclosure risk of all the algorithms fairly. For all these scenarios, we compute k' and the mean of all the k' values in each situation is presented in the third, fourth and fifth columns. Note that, whereas multivariate microaggregation is affected by the fact that the chosen attributes are correlated or not, this effect is not noticeable using Mic1D-k. Specifically, when the attributes in a group are not correlated, the information loss (SSE) using multivariate microaggregation tends to be increased since we are trying to collapse the records in a single value, using three independent attributes or dimensions. For instance, as it is illustrated in the first row of Table 5 (MDAV microaggregation with k equal to 5), the SSE value increases from 28.18 to 69.51 when the blocks are made using non correlated attributes . Nevertheless, this effect can be neglected with Mic1D-k since, thanks to the data preprocessing, the whole microaggregation process is performed on a single dimension (vector of values), the semantics of attributes are ignored and the effect caused by attribute correlations is avoided. For this reason, in Tables 4 and 5 SSE values are identical for the correlated and non correlated attribute blocking.

The results described in Tables 4 and 5 also show that, Mic1D-k achieves lower disclosure risk levels (larger values of k') than those achieved by multivariate microaggregation for similar information loss (SSE), especially when the attributes chosen come from different microaggregated groups (2G and 3G), which is the most common case. For instance, if we observe the k' values of MDAV microaggregation using the most 'private' configuration (k equal to 25 and using non correlated attributes) we can see that the resulting k' values where the intruder has access to attributes coming from more than one group (G2 equal to 9.22 and G3 equal to 1.91) are lower than using Mic1D-k with similar SSE value (G2 equal to 14.52 and G3 equal to 11.71). Note also that, when the intruder has access to the three attributes coming from a single microaggregated group, multivariate microaggregation configurations present k' values which are similar or, in some cases, even larger than those obtained by Mic1D-k (comparing cases with similar SSE). This is normal since such methods preserve the k-anonymity in this case. However, in the remaining scenarios (2G and 3G), that represent most of the cases, Mic1D-k achieves larger k' values than those obtained by multivariate microaggregation when similar SSE values are compared.

4.4   Method Comparison Using Generic Measures

We have repeated the experiments presented in [6] where a large variety of protection methods were compared using the Census microdata file based on the *score*, presented in Subsect. 2.3, to measure the results. We have computed the disclosure risk considering different scenarios ranging from the extreme case where the intruder knows only one attribute, to the opposite case where it knows all the attributes, as in [19]. Specifically, we have considered 512 different sets of attributes for each Mic1D-k and multivariate microaggregation parameterization. The total number of executions run in these experiments is 10752.

Table 6. Score k' of different microaggregation methods and parameterizations using the Census microdata file. Method-k corresponds to microaggregation using method Method (MDAV, PCP or Zscores) with initial anonymity value k.

	k	IL	DLD	ID	Score		k	IL	DLD	ID	Score
MDAV-k	5	31.67	34.11	69.70	41.79	MDAV-k	5	41.47	23.14	63.03	42.28
	25	51.14	11.70	54.65	42.16		25	51.47	6.56	49.07	39.64
	50	60.19	5.68	47.75	43.45		50	44.11	3.26	43.87	33.84
PCP-k	5	63.18	4.35	47.63	44.59	PCP-k	5	55.70	3.75	42.71	39.47
	25	57.31	2.23	38.53	38.85		25	78.39	2.02	34.81	48.40
	50	57.86	1.81	34.41	37.98		50	83.30	1.50	32.29	50.10
Zscores-k	5	59.64	12.13	57.71	47.28	Zscores-k	5	51.76	4.80	51.66	40.00
	25	79.90	6.96	52.86	54.90		25	86.58	2.37	47.19	55.68
	50	89.33	6.25	50.09	58.75		50	90.81	2.43	43.80	56.96
Mic1D-k	3000	22.17	35.95	54.16	33.61	Mic1D-k	3000	22.17	35.95	54.16	33.61
	4000	57.31	16.91	48.53	45.02		4000	57.31	16.91	48.53	45.02
	5000	82.44	4.94	32.22	50.51		5000	82.44	4.94	32.22	50.51
		Correlated attributes						Non-correlated attributes			

The first main conclusion extracted from the results presented in Table 6 is that the quality obtained by Mic1D-k is orthogonal to the degree of correlation between the attributes in a cluster. On the contrary, this correlation has a significant effect on the remaining techniques. For example, the best score for MDAV using correlated attributes is 41.79 while it is 33.84 using non-correlated attributes. In this experiments we test two extreme cases where all the attributes are correlated or none of them. In a real scenario, we would not be able to choose the attributes to be protected, and, as a consequence, we do not have any control on the correlation between them making the application of these multivariate microaggregation techniques less suitable than our proposal.

Also, the configuration process is simplified for Mic1D-k. While the other multivariate microaggregation methods must choose the best attribute blocking selection for clustering, which may be as difficult as the anonymization problem itself, we avoid this problem by replacing the attribute selection phase by a significantly less complex pre-processing phase.

Finally, Mic1D-k obtains the lowest score when k is equal to 3000 (33.61). MDAV algorithm also obtains low scores when non-correlated attributes are grouped together (k equal to 50, 33.84). However, as we have said before, using MDAV one has to decide which attributes are to be grouped together and this is not a straightforward decision.

5 Conclusions and Future Work

In this paper, we have presented a new type of microaggregation called *One Dimension microaggregation*. This microaggregation method significantly diminish the problem of attribute selection in multivariate microaggregation achieving in general a higher level of privacy than that obtained by three of the most well-known microaggregation algorithms. This is specially true as, from the attributes known by the intruder, the number of these coming from different microaggregation groups of multivariate microaggregation increases.

As future work, we plan to develop and implement a method for vector partitioning which considers the SSE value when the partitions are done so that we can reduce the SSE value of our method and, therefore, the information loss.

All in all, in this paper we show that microaggregation is a very useful method for the anonymization of complex records containing a large number of attributes, when it is combined with the data preprocessing proposed in our work.

Acknowledgments. This work is partially supported by the Ministry of Science and Technology of Spain under contract TIN2012-34557 and by the BSC-CNS Severo Ochoa program (SEV-2011-00067)

References

1. Adam, N.R., Wortmann, J.C.: Security-control for statistical databases: a comparative study. ACM Comput. Surv. **21**, 515–556 (1989)
2. Aggarwal, C.: On k-anonymity and the curse of dimensionality. In: Proceedings of the 31st International Conference on Very Large Databases, pp. 901–909 (2005)
3. Aggarwal, G., Feder, T., Kenthapadi, K., Khuller, S., Panigrahy, R., Thomas, D., Zhu, A.: Achieving anonymity via clustering. In: Proceedings of the 25th ACM Symposium on Principles of Databases Systems, pp. 153–162 (2006)
4. CASC: Computational Aspects of Statistical Confidentiality, European Project IST-2000-25069, http://neon.vb.cbs.nl/casc
5. Domingo-Ferrer, J., Torra, V.: Disclosure control methods and information loss for microdata, pp. 91–110 of [8] (2001)
6. Domingo-Ferrer, J., Torra, V.: A quantitative comparison of disclosure control methods for microdata, pp. 111–133 of [8] (2001)

7. Domingo-Ferrer, J., Mateo-Sanz, J.M.: Practical data-oriented microaggregation for statistical disclosure control. IEEE Trans. Knowl. Data Eng. **14**(1), 189–201 (2002)

8. Doyle, P., Lane, J., Theeuwes, J., Zayatz, L. (eds.): Confidentiality, Disclosure, and Data Access: Theory and Practical Applications for Statistical Agencies. Elsevier Science, New York (2001)

9. Felso, F., Theeuwes, J., Wagner, G.: Disclosure limitation in use: results of a survey, pp. 17–42 of [8] (2001)

10. Fung, B., Wang, K., Yu, P.: Top-down specialization for information and privacy preservation. In: Proceedings of the 21st IEEE International Conference on Data, Engineering, pp. 205–216 (2005)

11. Hansen, S., Mukherjee, S.: A polynomial algorithm for optimal univariate microaggregation. Trans. Knowl. Data Eng. **15**(4), 1043–1044 (2003)

12. Jolliffe, I.T.: Principal Component Analysis. Springer Series in Statistics. Springer, New York (2002). ISBN: 978-0-387-95442-4

13. Larsen, R.J., Marx, M.L.: An Introduction to Mathematical Statistics and Its Applications, 3rd edn. Prentice Hall, Upper Saddle River (2005). ISBN-10: 0131867938

14. Mateo-Sanz, J.M., Domingo-Ferrer, J.: A method for data-oriented multivariate microaggregation. In: Statistical Data Protection for Official Publications of the European, Communities, pp. 89–99

15. Murphy, P., M., Aha, D.W.: UCI Repository machine learning databases. http://www.ics.uci.edu/~mlearn/MLRepository.html, University of California, Department of Information and Computer Science, Irvine, CA (1994)

16. Nin, J., Torra, V.: Empirical analysis of database privacy using twofold integrals. In: Hao, Y., Liu, J., Wang, Y.-P., Cheung, Y., Yin, H., Jiao, L., Ma, J., Jiao, Y.-C. (eds.) CIS 2005, vol. 3801, pp. 1–8. LNAI. Springer, Heidelberg (2005)

17. Nin, J., Herranz, J., Torra, V.: On the disclosure risk of multivariate microaggregation. Data. Knowl. Eng. (DKE), Elsevier **67**(3), 399–412 (2008)

18. Nin, J., Herranz, J., Torra, V.: How to group attributes in multivariate microaggregation. Int. J. Uncertainty Fuzziness Knowl. Based Syst. **16**(1), 121–138 (2008)

19. Nin, J., Herranz, J., Torra, V.: Towards a more realistic disclosure risk assessment. In: Domingo-Ferrer, J., Saygın, Y. (eds.) PSD 2008, vol. 5262, pp. 152–165. LNCS. Springer, Heidelberg (2008)

20. Oganian, A., Domingo-Ferrer, J.: On the complexity of optimal microaggregation for statistical disclosure control. Stat. J. United Nations Econ. Comm. Europe **18**(4), 345–354 (2000)

21. Pagliuca, D., Seri, G.: Some results of individual ranking method on the system of enterprise accounts annual survey, Esprit SDC Project, Deliverable MI-3/D2 (1999)

22. Samarati, P., Sweeney, L.: Protecting privacy when disclosing information: k-anonymity and its enforcement through generalization and suppression. SRI International technical reports (1998)

23. Sande, G.: Exact and approximate methods for data directed microaggregation in one or more dimensions. Int. J. Unc. Fuzz. Knowl. Based Syst. **10**(5), 459–476 (2002)

24. Sebé, F., Domingo-Ferrer, J., Mateo-Sanz, J.M., Torra, V.: Post-masking optimization of the tradeoff between information loss and disclosure risk in masked microdata sets. In: Domingo-Ferrer, J. (ed.) Inference Control in Statistical Databases, vol. 2316, pp. 163–171. LNCS. Springer, Heidelberg (2002)

25. Sweeney, L.: Achieving k-anonymity privacy protection using generalization and suppression. Int. J. Unc. Fuzz. Knowl. Based Syst. **10**(5), 571–588 (2002)
26. Sweeney, L.: k-anonymity: a model for protecting privacy. Int. J. Unc. Fuzz. Knowl. Based Syst. **10**(5), 557–570 (2002)
27. U.S. Census Bureau, Data Extraction System. http://www.census.gov/ (1990)
28. Willenborg, L., Waal, T.: Elements of Statistical Diclosure Control. Lecture Notes in Statistics. Springer, New York (2001)

Author Index